THE
CARUS MATHEMATICAL MONOGRAPHS

Published by
THE MATHEMATICAL ASSOCIATION OF AMERICA

———

THE CARUS MATHEMATICAL MONOGRAPHS are an expression of the desire of Mrs. Mary Hegeler Carus, and of her son, Dr. Edward H. Carus, to contribute to the dissemination of mathematical knowledge by making accessible at nominal cost a series of expository presentations of the best thoughts and keenest researches in pure and applied mathematics. The publication of the first four of these monographs was made possible by a notable gift to the Mathematical Association of America by Mrs. Carus as sole trustee of the Edward C. Hegeler Trust Fund. The sales from these have resulted in the Carus Monograph Fund, and the Mathematical Association has used this as a revolving book fund to publish the succeeding monographs.

The expositions of the mathematical subjects which the monographs contain are set forth in a manner comprehensible not only to teachers and students specializing in mathematics, but also to scientific workers in other fields, and especially to the wide circle of thoughtful people who, having a moderate acquaintance with elementary mathematics, wish to extend their knowledge without prolonged and critical study of the mathematical journals and treatises. The scope of this series includes also historical and biographical monographs.

The following books in this series
have been published to date:

CALCULUS
OF VARIATIONS

By

GILBERT AMES BLISS

Professor of Mathematics, The University of Chicago

Published for

THE MATHEMATICAL ASSOCIATION OF AMERICA

by

THE OPEN COURT PUBLISHING COMPANY
LA SALLE, ILLINOIS

PREFACE

This book is the first of a series of monographs on mathematical subjects which are to be published under the auspices of the Mathematical Association of America and whose publication has been made possible by a very generous gift to the Association by Mrs. Mary Hegeler Carus as trustee for the Edward C. Hegeler Trust Fund. The purpose of the monographs is to make the essential features of various mathematical theories accessible and attractive to as many persons as possible who have an interest in mathematics but who may not be specialists in the particular theory presented, a purpose which Mrs. Carus has very appropriately described to be "the diffusion of mathematical and formal thought as contributory to exact knowledge and clear thinking, not only for mathematicians and teachers of mathematics but also for other scientists and the public at large."

The attainment of this end will not always be easy for authors who have long specialized in unraveling the intricacies of the domains in which their principal activities lie, and the clientele of readers which they may reasonably hope to interest will vary greatly with the subjects presented. It would obviously be unwise to regard this first attempt as in any final sense a model for the many monographs which it is hoped will follow. Later authors will doubtless profit much by the experiences of those who have written before, but varieties of subjects and types of readers to be addressed are likely to require an equally large variety of methods of presentation. It is possible that some monographs will be entirely descriptive or historical in character, others devoted to the treatment in detail of special mathematical questions which can be approached without elaborate prerequisite study, and still others of types not yet devised but which are certain to be suggested as the series progresses. One can readily foresee the beneficial influence which the monographs will have in encouraging and developing types of descriptive mathematical writing suited to the very laudable purposes for which the series has been inaugurated.

The theory to which the present monograph is devoted, the calculus of variations, is one whose development from the beginning has been interlaced with that of the differential and integral calculus. Without any knowledge of the calculus one can readily understand at least the geometrical or mechanical statements of many of the problems of the calculus of variations and the character of their solutions, as an examination of the chapters to follow will show. Thus if two points not in the same vertical line are given we may ask for the curve joining them down which a marble starting with a given initial velocity will roll from one point to the other in the shortest time. The solution is a piece of an inverted cycloid, and a cycloid is the curve described by a point on the rim of a wheel as the wheel rolls along the ground. Or if two points above a horizontal x-axis are given we may seek to find the curve which joins them and which when rotated around the x-axis generates a surface of revolution of minimum area. The solution curve will have sometimes one and sometimes the other of two forms. The first of these is the broken line consisting of the two perpendiculars from the points to the x-axis and the portion of the axis between them, in which case the minimum surface consists of two circular disks. The second is an arc of a catenary, and the form of a catenary is that which a chain naturally takes when suspended from two pegs. The surface generated in this latter case is the capstan-shaped surface assumed by a soap film suspended between two wire circles having a common axis.

The discovery and justification of the results which have just been described, apart from their simple statement, do require, however, acquaintance with the principles of the calculus, and in the following pages it is assumed that the reader has such an acquaintance. This should not deter others who may be interested from examining the introductions to the various chapters and the italicized theorems throughout the book, many of which should be perfectly intelligible to everyone. The only place where results not usually deduced in the ordinary calculus course are used is in the last chapter, where some properties of differential equations are required which have already been clearly illustrated in the three preceding chapters, and which are described in detail in the text.

In selecting material for presentation it seemed desirable to begin by studying special problems rather than the general theory. The first chapter of the book describes the historical setting out of which the theory of the calculus of variations grew, and the character of some of the simpler problems. The next three chapters are devoted to the development in detail of the known results for three special problems which illustrate in excellent fashion the essential characteristics of the general theory contained in Chapter V with which the book concludes.   The author was influenced in this selection by several considerations.   In the first place the theory of the special problems here presented requires only analysis of a concrete sort in which one is much aided by intuition while accumulating experiences which assist effectively in understanding later the notions of the general theory.   In the second place the theory of these problems, though well known, is scattered in various places in treatises and memoirs on the calculus of variations, and the presentation of it in collected form should therefore be useful as well as instructive.   Finally it is a fact that the modern theory of the calculus of variations has been presented for the most part in elaborate mathematical treatises and is not readily accessible except to the specialist.   The elementary discussions of the theory, in the larger more general treatises on analysis and also in separate form, usually lay their emphasis upon the deduction of the differential equations of the minimizing curves for various types of problems, and relatively little upon other aspects of the theory.   It is doubtless partly for this reason that in applied mathematics much more use has hitherto been made of these differential equations and their solutions than of the further properties of minimizing curves, though it is well known that in many cases these further properties are closely related to interesting conditions for stability in associated problems of mechanics.

Such are the reasons why it seemed desirable to the author to present in this book the theory of special problems with some completeness, even if limitations of space should permit only a few of them to be discussed. It must be admitted that in the literature of the calculus of variations there are not many particular cases to which the general theory has been thoroughly applied. The assembling of as many such problems as possible and the completion of others would be a work of great usefulness and interest.

At the end of Chapter V is a list of the books on the calculus of variations with a few other references of importance for the topics considered in the text. The notes, indicated serially in the text by superscripts, follow this list of references.

G. A. BLISS

THE UNIVERSITY OF CHICAGO
    October, 1924

TABLE OF CONTENTS

CHAPTER I

TYPICAL PROBLEMS OF THE CALCULUS OF VARIATIONS

1. *The invention of the calculus.* When the student of mathematics pauses to look back upon the achievements of mathematicians of the past he must be impressed with the fact that the seventeenth century was a most important epoch in the development of modern mathematical analysis, since to the mathematicians of that period we owe the invention of the differential and integral calculus.   At first the calculus theory, if indeed at that time it could be called such, consisted of isolated and somewhat crude methods of solving special problems.   In the domain of what we now call the integral calculus, for example, an Italian mathematician named Cavalieri (1598–1647) devised early in the seventeenth century a summation process, called the method of indivisibles, by means of which he was able to calculate correctly many areas and volumes.   His justification of his device was so incomplete logically, however, that even in those relatively uncritical times his contemporaries were doubtful and dissatisfied.   Somewhat later two French mathematicians, Roberval (1602–75) and Pascal (1623–62), and the Englishman Wallis (1616–1703), improved the method and made it more like the summation processes of the integral calculus of today.   In the case of the differential calculus we find that before the final quarter of the seventeenth century Descartes (1596–1650), Roberval, and Fermat (1601–65) in France, and Barrow

(1630–77) in England, all had methods of constructing tangents to curves which were pointing the way toward the solution of the fundamental problem of the differential calculus as we formulate it today, namely, that of determining the slope of the tangent at a point of a curve.

At this important stage there appeared upon the scene two men of extraordinary mathematical insight, Newton (1642–1727) in England, and Leibniz (1646–1716) in Germany, who from two somewhat different standpoints carried forward the theory and applications of the calculus with great strides. It is a mistake, though we often find it an easy convenience, to regard these two great thinkers as having invented the calculus out of a clear sky. Newton was in fact a close student of the work of Wallis, and a pupil of Barrow whom he succeeded as professor of mathematics at Cambridge, while we know that Leibniz visited Paris and London early in his career and that he there became acquainted with the most advanced mathematics of his day. No one could successfully contest the fact, however, that these two men were the most able spokesmen and investigators of the seventeenth-century school of mathematicians to which we owe the gradual evolution of the calculus.

In spite of the great abilities of Newton and Leibniz the underlying principles of the calculus as exposed by them seem to us from our modern viewpoint, as indeed to their contemporaries and immediate successors, somewhat vague and confusing. The difficulty lies in the lack of clearness at that early time, and for more than a century thereafter, in the conceptions of infinitesimals and limits upon which the calculus rests, a difficulty which has been overcome only by the systematic study of the theory of limits inaugurated by Cauchy (1789–1857) and continued by Weierstrass (1815–97), Riemann (1826–66), and many others.

2. *Maxima and minima.* Among the earliest problems which attracted the attention of students of the calculus were those which require the determination of a maximum or a minimum. Fermat had devised as early as 1629 a procedure applicable to such problems, depending upon principles which in essence, though not in notation, were those of the modern differential calculus. Somewhat nearer to the type of reasoning now in common use are the methods which Newton and Leibniz applied to the determination of maxima and minima, methods which are also characteristic of their two conceptions of the fundamental principles of the differential calculus. Newton argued, in a paper written in 1671 but first published in 1736, that a variable is increasing when its rate of change is positive, and decreasing when its rate is negative, so that at a maximum or a minimum the rate must be zero. Leibniz, on the other hand, in a paper which he published in 1684, conceived the problem geometrically. At a maximum or a minimum point of a curve the tangent must be horizontal and the slope of the tangent zero.

At the present time we know well that from a purely analytical standpoint these two methods are identical. The derivative

$$(1) \qquad f'(x) = \lim_{\Delta x \to 0} \frac{f(x + \Delta x) - f(x)}{\Delta x}$$

of a function $f(x)$ represents both the rate of change of $f(x)$ with respect to x and the slope of the tangent at a point on the graph of $f(x)$. For in the first place the fraction in the second member of equation (1) is the average rate of change of $f(x)$ with respect to x on the interval from x to $x+\Delta x$, and its limit as the interval is shortened is therefore rightly called the rate of change

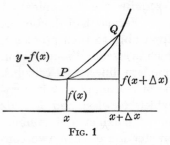

Fig. 1

of $f(x)$ at the initial value x of the interval. In the second place this same quotient is the slope of the secant PQ in Figure 1, and its limit is the slope of the tangent at P. Thus by the reasoning of either Newton or Leibnitz we know that the maxima and minima of $f(x)$ occur at the values of x where the derivative $f'(x)$ is zero.

It was not easy for the seventeenth-century mathematician to deduce this simple criterion that the derivative $f'(x)$ must vanish at a maximum or a minimum of $f(x)$. He was immersed in the study of special problems rather than general theories, and had no well-established limiting processes or calculus notations to assist him. It was still more difficult for him to advance one step farther to the realization of the significance of the second derivative $f''(x)$ in distinguishing between maximum and minimum values. Leibniz in his paper of 1684 was the first to give the criterion. In present-day parlance we say that $f'(a)=0$, $f''(a)\geqq 0$ are necessary conditions for the value $f(a)$ to be a minimum, while the conditions $f'(a)=0$, $f''(a)>0$ are sufficient to insure a minimum. For a maximum the inequality signs must be changed in sense.

It will be noted that the conditions just stated as necessary for a minimum are not identical with those which are sufficient. We shall see in Chapter V that a similar undesirable and much more baffling discrepancy occurs in the calculus of variations. For the simple problem of minimizing a function $f(x)$ the doubtful intermediate case when $f'(a)$ and $f''(a)$ are both zero was discussed by Maclaurin (1698–1746) who showed how higher derivatives may be used to obtain criteria which are both necessary and sufficient. For the calculus of variations the corresponding problem offers great difficulty and has never been completely solved.

3. *Two problems of the calculus of variations which may be simply formulated.* When one realizes the difficulty with which the late seventeenth-century school of mathematicians established the first fundamental principles of the calculus and their applications to such elementary problems in maxima and minima as the one which has just been described, it is remarkable that they should have conceived or attempted to solve with their relatively crude analytical machinery the far more difficult maximum and minimum problems of the calculus of variations which were at first proposed. It is an interesting fact that these early problems were not by any means the least complicated ones of the calculus of variations, and we shall do well therefore to introduce ourselves to the subject by looking first at two others which are easier to describe to one who has not already amused himself by browsing in this domain of mathematics.

The simplest of all the problems of the calculus of variations is doubtless that of determining the shortest arc joining two given points. The co-ordinates of these points will always be denoted by (x_1, y_1) and (x_2, y_2) and we may designate the points themselves when convenient simply by the numerals 1 and 2. If the equation of an arc is taken in the form

$$y = y(x) \qquad (x_1 \leq x \leq x_2)$$

then the conditions that it shall pass through the two given points are

$$(2) \qquad y(x_1) = y_1, \qquad y(x_2) = y_2,$$

and we know from the calculus that the length of the arc is given by the integral

$$I = \int_{x_1}^{x_2} \sqrt{1 + y'^2}\, dx,$$

where in the evaluation of the integral y' is to be replaced by the derivative $y'(x)$ of the function $y(x)$ defining the arc. There is an infinity of curves $y = y(x)$ joining the points 1 and 2. The problem of finding the shortest one is equivalent analytically to that of finding in the class of functions $y(x)$ satisfying the conditions (2) one which makes the integral I a minimum.

In the more elementary minimum problem of Section 2 above a function $f(x)$ is given and a value $x = a$ is sought for which the corresponding value $f(a)$ is a minimum. In the shortest-distance problem the integral I takes the place of $f(x)$, and instead of a value $x = a$ making $f(a)$ a minimum we seek to find an arc E_{12} joining the points 1 and 2 which shall minimize I. The analogy between the two problems is more perspicuous if we think of the length I as a function $I(E_{12})$ whose value is uniquely determined when the arc E_{12} is given, just as $f(x)$ in the former case was a function of the variable x.

There is a second problem of the calculus of variations, of a geometrical-mechanical type, which the principles of the calculus readily enable us to express also in analytic form. When a wire circle is dipped in a soap solution and withdrawn, a circular disk of soap film bounded by the circle is formed. If a second smaller circle is made to touch this disk and then moved away the two circles will be joined by a surface of film which is a surface of revolution in the particular case when the circles are parallel and have their centers on the same axis perpendicular to their planes. The form of this surface is shown in Figure 2. It is provable by the principles of mechanics, as one may readily surmise intuitively from the elastic properties of a soap film, that the surface of revolution so formed must be one of minimum area, and the problem of determining the shape of the film is equivalent therefore to that of determining such a minimum surface of revolution passing through two circles whose relative positions are supposed to be given as indicated in the figure.

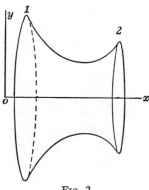

Fig. 2

In order to phrase this problem analytically let the common axis of the two circles be taken as the x-axis, and let the points where the circles intersect an xy-plane through that axis be 1 and 2. If the meridian curve of the surface in the xy-plane has an equation $y=y(x)$ then the calculus formula for the area of the surface is 2π times the value of the integral

$$I=\int_{x_1}^{x_2} y\sqrt{1+y'^2}\,dx \;.$$

The problem of determining the form of the soap film surface between the two circles is analytically that of finding in the class of arcs $y=y(x)$ whose ends are at the points 1 and 2 one which minimizes the last-written integral I.

4. *The problem of Newton.* It was remarked above that the earliest problems of the calculus of variations were not by any means the simplest. In his *Principia* (1686)[1] Newton states without proof certain conditions which must be satisfied by a surface of revolution which is so formed that it will encounter a minimum resistance when moved in the direction of its axis through a resisting medium. A particular case of the problem of finding such a surface is the well-known one of determining the form of a projectile which for a specified initial velocity will give the longest range. In practical ballistics it turns out that one of the most difficult parts of the investigation of this question lies in the experimental determination of the retardation law for bodies moving in the air at high rates of speed. Newton assumed a relatively simple law of resistance for bodies moving in a resisting medium which does not agree well with our experience with bodies moving in the air, but on the basis of which he was able to find a condition characterizing the meridian curves of the surfaces of revolution which encounter minimum

resistance. From a letter written by Newton to Professor David Gregory, probably in 1694, Bolza has reconstructed in most interesting fashion the arguments which Newton used in attaining his results.[1]

It is sufficient for the purposes of this introductory chapter to say that when the surface is generated by rotating about the x-axis an arc with an equation of the form $y = y(x)$ the resistance experienced by the surface when moved in the direction of the x-axis will, except for a constant factor, be

$$I = \int_{x_1}^{x_2} \frac{yy'^3}{1+y'^2}\, dx .$$

Newton's problem in analytical form is then that of determining among all the arcs $y = y(x)$ joining two given points 1 and 2 one which makes this integral a minimum. We could equally well of course ask to determine the curve so that the resistance should be a maximum. If the law of resistance of Newton is replaced by another the methods which we now know of attacking the problem will still be applicable, though the results may be different, as a number of writers have shown.

5. *The brachistochrone problem.* Newton's problem, published in the *Principia* in 1686, lay apparently unnoticed for more than a decade before a new interest aroused by a second and more famous problem of the calculus of variations caused it to be studied again. It is not surprising that this happened because Newton's description of his results is very informal and concise. He gave no hint of a larger class of similar questions, and no suggestion of a method of solution which might have been applicable to such a class. To discover the beginnings

of active research in the calculus of variations we must turn therefore to other writers.

In the period which followed the discovery and publication of the calculus methods of Newton and Leibnitz two of the most prominent and successful researchers in the new analysis were the Swiss mathematicians James Bernoulli (1654–1705), professor of mathematics at the University of Basle, and his brother John (1667–1748). The younger brother was a student of the elder, and among those students he was in later years by far the most distinguished on account of his varied and successful researches. He studied with James until the year 1690 when he forsook Basle for travel and the study of mathematics in France. Shortly after his return he accepted in 1695 a professorship at the University of Groningen, and in 1705, upon the death of James Bernoulli, he returned to Basle to spend the remainder of his life as professor of mathematics in his native city.

In the years just preceding 1695 a rivalry sprang up between the two brothers the reason for which is not well understood. It was at times amusingly undignified, and from the scientific standpoint unjustifiable, since both brothers were with somewhat different temperaments remarkably able and worthy of respect. Whatever may have been the cause of their dissension it is at any rate true that the friction between them gave an unusual impetus and zest to the beginnings of the calculus of variations. In June, 1696, John Bernoulli proposed his now famous brachistochrone problem, and publicly incited the mathematicians of the world to give it their consideration, according to a custom which was common at the time.[2] We know that the problem

aroused great interest and that Newton, Leibniz, and l'Hospital (1661–1704), besides the brothers Bernoulli, found the correct solution.

The problem of the brachistochrone ($\beta\rho\alpha\chi\iota\sigma\tau\sigma\varsigma$ = shortest, $\chi\rho\sigma\nu\sigma\varsigma$ = time) is that of determining a path down which a particle will fall from one given point to another in the shortest time. Let the y-axis for convenience be taken vertically downward, as in Figure 3, the two fixed points being 1 and 2. The

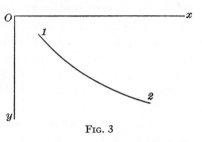

FIG. 3

initial velocity v_1 at the point 1 is supposed to be given. In Chapter III we shall see that for a curve defined by an equation of the form $y = y(x)$ the time of descent from 1 to 2 is $1/\sqrt{2g}$ times the value of the integral

$$I = \int_{x_1}^{x_2} \sqrt{\frac{1+y'^2}{y-a}} \, dx \, ,$$

where g is the gravitational constant and a has the constant value $a = y_1 - v_1^2/2g$. The problem of the brachistochrone is then to find among the curves $y = y(x)$ which pass through two points 1 and 2 one which minimizes the integral I.

The only discussions of the problem which were published in full in response to John Bernoulli's invitation were those of the Bernoulli brothers themselves,[3] in May, 1697, and they are in many respects characteristic of their authors. John's paper is to this day most elegant and satisfactory reading. He saw that the curve of quickest descent is identical with the path of a ray of light in a medium with a suitably selected variable index of refraction, and a known property of such paths enabled him to attain very quickly and easily a solution. His method can be applied, however, to only a restricted class of similar questions. The solution of James was more laborious, and to us much less attractive, since it was couched in the language of the relatively clumsy geometrical analysis which preceded the invention of the calculus and which was commonly used for some time thereafter. But his method was a more general one than that of his younger brother and was the first step in a long series of researches which has led to the theory of the calculus of variations as we know it today.

At the close of his paper James invited mathematicians in general to consider a much more difficult problem of the calculus of variations which he had devised, and he offered to John in particular a money prize of fifty ducats for a satisfactory solution. As it turned out, however, the ducats were saved, for although John claimed to have done so he did not as a matter of fact succeed in his attacks upon the problem, and after a rather bitter discussion which dragged on for a number of years James finally published his own solution in 1701. The two papers of 1697 and 1701 of James Bernoulli were the starting-point for the researches of Euler (1707–83), a native of Basle and pupil of John Bernoulli, one of the greatest of the world's mathematicians. It is to Euler that we owe the first important result in the modern theory of the calculus of variations, as we shall see in later chapters.

It is fair to say that the theory of the calculus of

variations had its beginning in the interesting brachisto-chrone problem of John Bernoulli. One should not infer from this remark that no problems of the calculus of variations were known earlier, for we have seen already that Newton had proposed such a problem and described a characteristic property of its solution. Furthermore the brachistochrone problem itself was more or less definitely in the mind of Galileo (1564–1642) in 1630 and 1638[4] when he compared the time of fall of a particle along an arc of a vertical circle with those along polygons inscribed in the arc. He seems to conclude that the time of descent on a circular arc is shorter than the times on all other paths joining its end-points, but his proof does not justify this result. Nowadays we know that the solution curve is neither a circle nor a straight line but a cycloid, as will be proved in Chapter III. A still older problem of the calculus of variations is the iso-perimetric problem of the ancient Greeks described in Section 7 below. None of these, however, could rightly be regarded as the starting-point of the theory of the calculus of variations, for in the early references to them there were no indications of other problems of similar type, or of methods of solution possessing generality of application.

6. *A more general problem.* With the exception of the integral in Newton's problem those which have been mentioned in the preceding sections all have the form

$$(3) \qquad I = \int_{x_1}^{x_2} n(x, y) \sqrt{1 + y'^2} \, dx \, ,$$

and we might propose to ourselves to find among the curves $y = y(x)$ joining two given points one which minimizes this integral I. This problem also has a physical interpretation. For suppose that in a plane transparent medium the velocity of light varies from point to point, and that at an arbitrary point (x, y) it has the value $v(x, y)$. The index of refraction at that point has by definition the value $n(x, y) = c/v(x, y)$, where c is a constant, and the time dt taken by a disturbance to travel along an arc of length ds through the point (x, y) with the velocity $v(x, y)$ is approximately

$$dt = \frac{ds}{v(x, y)} = \frac{1}{c} n(x, y) \sqrt{1+y'^2}\, dx .$$

We see readily by an integration that the integral I is proportional to the time taken by a disturbance to traverse the arc $y = y(x)$ joining the two given points 1 and 2. Now it has been verified physically that the path of a ray of light in a medium in which the velocity of light varies from point to point is always one on which the time-integral I is, for short arcs at least, a minimum, so that our problem of minimizing I is that of determining the paths of rays of light in a plane medium whose variable index of refraction is $n(x, y)$.

John Bernoulli noted that the time of descent of a particle down a curve $y = y(x)$, and the time of passage of a ray along the same curve in a medium with the index of refraction $n(x, y) = c/\sqrt{y-a}$, are, except for a constant factor, given by the same integral (3) with this index substituted. He knew furthermore that when a ray of light passes from one medium to another the sines of the angles of incidence and refraction at the bounding surface are proportional to the indices of refraction in the two media, and by thinking of his medium as made up of very thin horizontal layers with different indices he was able to deduce the form of the curve of quickest descent.

The integral (3) still does not include that of Newton's problem as a special case, though it is general enough to so include most of the classical special problems of the calculus of variations in the plane. It will be quite as easy for us, however, to consider an integral of the form

$$(4) \qquad I = \int_{x_1}^{x_2} f(x, y, y')dx ,$$

having an integrand which is an arbitrary function of the three variables x, y, y', as we shall do in Chapter V. Among all the arcs $y = y(x)$ joining two given points 1 and 2 we shall seek one which minimizes the integral (4). This is a problem of sufficient generality to include all of those hitherto stated as special cases.

7. *Other problems of the calculus of variations.* It would be a mistake to infer that the category of questions to which the calculus of variations is devoted is exhausted even by the quite general problem proposed in the last section. We can vary the problem there described by seeking a minimizing curve among those joining a fixed point and a fixed curve, or two fixed curves, instead of two fixed points, or in many other ways.

The famous old isoperimetric problem of the ancients was that of finding a simply closed curve of given length which incloses the largest area. The solution is a circle, though it is not any too easy to prove that this is so. Analytically the problem may be formulated as that of finding an arc with equations in the parametric form

$$x = x(t) , \qquad y = y(t) \qquad (t_1 \leqq t \leqq t_2)$$

satisfying the conditions

$$x(t_1) = x(t_2) , \qquad y(t_1) = y(t_2)$$

but not otherwise intersecting itself, giving the length integral

$$\int_{t_1}^{t_2} \sqrt{x'^2 + y'^2}\, dt$$

a fixed value l, and maximizing the area integral

$$\tfrac{1}{2} \int_{t_1}^{t_2} (xy' - x'y)\, dt .$$

The problems of the calculus of variations for which one or more integrals are to be given fixed values, while another is to be made a minimum or maximum, are called, after this one, isoperimetric problems. The problem proposed by James Bernoulli in 1697 was the earliest isoperimetric problem after that of the ancient Greeks.

It will not be possible for us in the limited space of the following pages to examine in detail more than the simpler non-isoperimetric problems, though there are many other types besides those which have already been mentioned.

The theory of the calculus of variations has been extensively developed but not so widely applied to special cases, very few of the particular problems having been exhaustively investigated. In the following Chapters II–IV three of the special problems mentioned in the preceding pages which have been studied in detail will be discussed, and in Chapter V some of the results for the more general problem formulated in Section 6 are collected, with a brief historical sketch of the progress of the theory from the time of the Bernoullis to the present.

CHAPTER II

SHORTEST DISTANCES

8. *The shortest arc joining two points*. Problems of
determining shortest distances furnish a useful elementary
introduction to the theory of the calculus of variations
because the properties characterizing their solutions are
familiar ones which illustrate very well many of the general principles common to all of the problems suggested in
the preceding chapter. If we can for the moment eradi-
cate from our minds all that we know about straight lines
and shortest distances we shall have the pleasure of re-
discovering well-known theorems by methods which will
be helpful in solving more complicated problems.

Let us begin with the simplest case of all, the problem
of determining the shortest arc joining two given points.
The integral to be minimized, which we have already
seen on page 6 of the preceding chapter, may be written
in the form

$$(1) \qquad I = \int_{x_1}^{x_2} f(y') dx$$

if we use the notation $f(y') = (1 + y'^2)^{\frac{1}{2}}$, and the arcs
$y = y(x)$ $(x_1 \leqq x \leqq x_2)$ whose
lengths are to be compared
with each other will always
be understood to be continuous and to consist of a finite

Fig. 4

number of arcs on each of which the tangent turns continuously, as indicated in Figure 4. Analytically this means

that on the interval $x_1 \leqq x \leqq x_2$ the function $y(x)$ is continuous, and that the interval can be subdivided into parts on each of which $y(x)$ has a continuous derivative.   Let us agree to call such functions *admissible functions* and the arcs which they define admissible arcs.   Our problem is then to find among all admissible arcs joining two given points 1 and 2 one which makes the integral I a minimum.

9. *A first necessary condition.*   Let it be granted that a particular admissible arc E_{12} with the equation

$$y = y(x) \qquad (x_1 \leqq x \leqq x_2)$$

furnishes the solution of our problem, and let us then seek to find the properties which distinguish it from the other admissible arcs joining points 1 and 2.   If we select arbitrarily an admissible function $\eta(x)$ satisfying the conditions $\eta(x_1) = \eta(x_2) = 0$, the equation

$$(2) \qquad y = y(x) + a\,\eta(x) \qquad (x_1 \leqq x \leqq x_2) ,$$

involving the arbitrary constant a, represents a one-parameter family of curves which includes the arc E_{12} for the special value $a = 0$, and all of the curves of the family pass through the end-points 1 and 2 of E_{12}. The value of the integral I taken along an arc of the family depends upon the value of a and may be represented by the symbol

$$(3) \qquad I(a) = \int_{x_1}^{x_2} f(y' + a\eta')dx .$$

Along the initial arc E_{12} the integral has the value $I(0)$, and if this is to be a minimum when compared with the values of the integral along all other admissible arcs joining 1 with 2 it must in particular be a minimum when compared with the values $I(a)$ along the arcs of the family (2). Hence according to the criterion for a minimum of a function given on page 4 of the last chapter we must have $I'(0) = 0$.

It should perhaps be emphasized here that the method of the calculus of variations, as it has been developed in the past, consists essentially of three parts; first, the deduction of necessary conditions which characterize a minimizing arc; second, the proof that these conditions, or others obtained from them by slight modifications, are sufficient to insure the minimum sought; and third, the search for an arc which satisfies the sufficient conditions. For the deduction of necessary conditions the value of the integral I along the minimizing arc can be compared with its values along any special admissible arcs which may be convenient for the purposes of the proof in question, for example along those of the family (2) described above, but the sufficiency proofs must be made with respect to all admissible arcs joining the points 1 and 2. The third part of the problem, the determination of an arc satisfying the sufficient conditions, is frequently the most difficult of all, and is the part for which fewest methods of a general character are known. For shortest-distance problems fortunately this determination is usually easy.

By differentiating the expression (3) with respect to a and then setting $a = 0$ the value of $I'(0)$ is readily seen to be

$$(4) \qquad I'(0) = \int_{x_1}^{x_2} f_{y'}\, \eta'dx ,$$

where for convenience we use the notation $f_{y'}$ for the derivative of the integrand $f(y')$ with respect to y'. It will always be understood that the argument in f and its derivatives is the function $y'(x)$ belonging to the arc E_{12} unless some other is expressly indicated, as is done, for example, in the formula (3).

What now are the conclusions which can be drawn from the necessity of the condition $I'(0) = 0$? The answer to this question is to be found in the lemma of the following section which will be frequently applied in later chapters as well as in the solution of the shortest-distance problems to which this chapter is devoted.

10. *A fundamental lemma.*[5] In the integrand of the integral (4) the coefficient of η' is really a function of x, since the derivative $f_{y'}$ contains as its argument the slope $y'(x)$ of the arc E_{12}, and we may denote this coefficient by $M(x)$. It should be noted that the function $M(x)$ is continuous except possibly at the values of x defining the corners of the arc E_{12} where the slope $y'(x)$ changes abruptly. At those points of the curve it has two values, one corresponding to the backward and one to the forward slope. The lemma which we wish to prove is then as follows:

FUNDAMENTAL LEMMA. *Let $M(x)$ be a function of the kind described above, continuous on the interval $x_1 \leqq x \leqq x_2$ or else such that the interval can be subdivided into a finite number of parts on each of which $M(x)$ is continuous. If the integral*

$$\int_{x_1}^{x_2} M(x)\eta'(x)dx$$

vanishes for every admissible function $\eta(x)$ such that $\eta(x_1) = \eta(x_2) = 0$, then $M(x)$ is necessarily a constant.

To see that this is so we note first that the vanishing of the integral of the lemma implies also the equation

$$(5) \qquad \int_{x_1}^{x_2} [M(x)-C]\eta'(x)dx=0$$

for every constant C, since all the functions $\eta(x)$ to be considered have $\eta(x_1)=\eta(x_2)=0$. The particular function $\eta(x)$ defined by the equation

$$(6) \qquad \eta(x)=\int_{x_1}^{x} M(x)dx-C(x-x_1)$$

evidently has the value zero at $x=x_1$, and it will vanish again at $x=x_2$ if, as we shall suppose, C is the constant value satisfying the condition

$$0=\int_{x_1}^{x_2} M(x)dx-C(x_2-x_1) .$$

The function $\eta(x)$ defined by equation (6) with this value of C inserted is now one of those which must satisfy equation (5). Its derivative is $\eta'(x)=M(x)-C$ except at points where $M(x)$ is discontinuous, since the derivative of an integral with respect to its upper limit is the value of the integrand at that limit whenever the integrand is continuous at the limit. For the special function $\eta(x)$, therefore, equation (5) takes the form

$$\int_{x_1}^{x_2} [M(x)-C]^2dx=0$$

and our lemma is an immediate consequence since this equation can be true only if $M(x)\equiv C$.

11. *Proof that the straight line is shortest.* In the equation $y=y(x)+a\,\eta(x)$ of the family of curves passing through the points 1 and 2 the function $\eta(x)$ was entirely

arbitrary except for the restrictions that it should be admissible and satisfy the relations $\eta(x_1)=\eta(x_2)=0$, and we have seen that the expression (4) for $I'(0)$ must vanish for every such family. The lemma of the preceding section is therefore applicable and it tells us that along the minimizing arc E_{12} an equation

$$f_{y'}=\frac{y'}{\sqrt{1+y'^2}}=C$$

must hold, where C is a constant. If we solve this equation for y' we see that y' is also a constant along E_{12} and that the only possible minimizing arc is therefore a single straight-line segment without corners joining the point 1 with the point 2.

The property just deduced for the shortest arc has so far only been proved to be necessary for a minimum. We have not yet demonstrated conclusively that the straight-line segment E_{12} joining 1 and 2 is actually shorter than every other admissible arc joining these points. In order to actually establish this fact let us now use $\eta(x)$ to denote the increment which must be added to the ordinate of E_{12} at the value x in order to get the ordinate of an arbitrarily selected admissible arc C_{12} joining 1 with 2, so that the equation of C_{12} will be

$$y=y(x)+\eta(x) \qquad (x_1\leqq x\leqq x_2) .$$

The difference between the lengths of C_{12} and E_{12} can now be expressed with the help of Taylor's formula in the form

$$I(C_{12})-I(E_{12})=\int_{x_1}^{x_2}\{f(y'+\eta')-f(y')\}dx$$

$$=\int_{x_1}^{x_2}f_{y'}\eta'dx+\tfrac{1}{2}\int_{x_1}^{x_2}f_{y'y'}(y'+\theta\eta')\eta'^2dx ,$$

where $I(C_{12})$ and $I(E_{12})$ are the values of the integral I along the two arcs; $f_{y'y'}$ is the second derivative of the function f with respect to y'; and θ is the value between 0 and 1 introduced by Taylor's formula. The next to last integral vanishes since f_y is a constant along E_{12} and since the difference $\eta(x)$ of the ordinates of two arcs C_{12} and E_{12} with the same end-points must vanish at x_1 and x_2. Furthermore the last integral is never negative since the second derivative $f_{y'y'} = 1/(1+y'^2)^{\frac{3}{2}}$ is always positive. We see therefore that $I(C_{12}) - I(E_{12})$ is greater than zero unless $\eta'(x)$ vanishes identically, in which case $\eta(x)$ itself would have everywhere the constant value zero which it has at x_1 and x_2, and C_{12} would coincide with E_{12}.

It has been proved therefore that the shortest arc from the point 1 to the point 2 is necessarily the straight-line segment joining those points, and that this segment is actually shorter than every other admissible arc with the same end-points.

One should notice the rôle which the positive sign of the derivative $f_{y'y'}$ has played in the determination of the minimum property. If the sign of this derivative had been negative the difference $I(C_{12}) - I(E_{12})$ would have been negative and $I(E_{12})$ would have been a maximum instead of a minimum. This is an analogue of the criterion mentioned on page 4 for the simpler theory of maxima and minima of functions of a single variable.

12. *Two important auxiliary formulas.* The type of proof used in the preceding section to show that the straight line joining 1 with 2 is shorter than every other admissible arc joining those two points is a very special one, not applicable in general to problems of the calculus

of variations whose integrals I have integrands containing one or both of the variables x and y as well as y'. It will be well worth while, therefore, to consider a second form of proof which will extend somewhat the results already found for the problem of finding the shortest distance between two points, and which will be applicable not

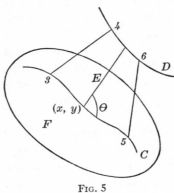

FIG. 5

only to the problems of shortest distances considered in this chapter but also to those which we shall study later.

We shall need first of all two special cases of more general formulas which are frequently applied in succeeding pages. Let E_{34} be a straight-line segment of variable length which moves so that its end-points describe simultaneously the two curves C and D shown in Figure 5, and let the equations of these curves in parametric form be

$$(C) \quad x=x_3(t), \; y=y_3(t) \, ,$$
$$(D) \quad x=x_4(t), \; y=y_4(t) \, .$$

The length

$$I=\sqrt{(x_4-x_3)^2+(y_4-y_3)^2}$$

of the segment E_{34} has the differential

$$dI=\frac{(x_4-x_3)(dx_4-dx_3)+(y_4-y_3)(dy_4-dy_3)}{\sqrt{(x_4-x_3)^2+(y_4-y_3)^2}} \, .$$

When the notation $p = (y_4 - y_3)/(x_4 - x_3)$ is used to denote the slope of the line E_{34} this result may be expressed in the more convenient formula of the following theorem:

If a straight-line segment E_{34} moves so that its end-points 3 and 4 describe simultaneously two curves C and D, as shown in Figure 5, then the length I of E_{34} has the differential

$$(7) \qquad dI(E_{34}) = \frac{dx + p\,dy}{\sqrt{1 + p^2}} \Big|_3^4$$

where the vertical bar indicates that the value of the preceding expression at the point 3 is to be subtracted from its value at the point 4. In this formula the differentials dx, dy at the points 3 and 4 are those belonging to C and D, while p is the slope of the segment E_{34}.

We shall need frequently to integrate the expression in the second member of equation (7) along curves such as C and D. This is evidently justifiable along C, for example, since the slope $p = (y_4 - y_3)/(x_4 - x_3)$ is a function of t and since the differentials dx, dy can be calculated in terms of t and dt from the equations of C, so that the expression takes the form of a function of t multiplied by dt. The integral I^* defined by the formula

$$I^* = \int \frac{dx + p\,dy}{\sqrt{1 + p^2}}$$

will also be well defined along an arbitrary curve C when p is a function of x and y, provided that we agree to calculate the value of I^* by substituting for x, y, dx, dy the expressions for these variables in terms of t and dt obtained from the parametric equations of C.

Let t_3 and t_5 be two parameter values which define points 3 and 5 on C, and which at the same time define two corresponding points 4 and 6 on D, as in Figure 5. If we integrate the formula (7) with respect to t from t_3 to t_5 and use the notation I^* just introduced for the integral of its second member, we find as a further result:

The difference of the lengths $I(E_{34})$ and $I(E_{56})$ of the moving segment in two positions E_{34} and E_{56} is given by the formula

$$(8) \qquad I(E_{56}) - I(E_{34}) = I^*(D_{46}) - I^*(C_{35}) \ .$$

This and the formula (7) are the two important ones for which we have been seeking. It is evident that they will still hold in even simpler form when one of the curves C or D degenerates into a point, since along such a degenerate curve the differentials dx and dy are zero.

The integrand of the integral I^* has a simple geometrical interpretation at the points of the curve C along which it is taken. At the point (x, y) of the curve C in Figure 5, for example, the angles between the x-axis and the tangents to C and E have, respectively, the cosines and sines

$$\frac{x'}{\sqrt{x'^2+y'^2}}, \quad \frac{y'}{\sqrt{x'^2+y'^2}}; \quad \frac{1}{\sqrt{1+p^2}}, \quad \frac{p}{\sqrt{1+p^2}}.$$

Since the angle θ between these tangents, and the element of length ds on C, are defined by the equations

$$(9) \qquad \cos \theta = \frac{x'+py'}{\sqrt{(1+p^2)(x'^2+y'^2)}}, \quad ds = \sqrt{x'^2+y'^2}\,dt$$

it follows that the integral I^* can also be expressed in the convenient form

$$(10) \qquad I^* = \int \frac{dx+p\,dy}{\sqrt{1+p^2}} = \int \cos \theta \ ds \ .$$

13. *The notion of a field and a second sufficiency proof.*
We have seen on page 19 that necessary conditions on the
shortest arc may be deduced by comparing it with other
admissible arcs of special types, but that a particular
line can be proved to be actually the shortest only by
comparing it with all of the admissible arcs joining the
same two end-points. The sufficiency proof of this
section is valid not only for the arcs which we have named
admissible but also for arcs with equations in the parametric form

(11) $$x = x(t), \qquad y = y(t) \qquad (t_3 \leqq t \leqq t_5) .$$

We suppose always that the functions $x(t)$ and $y(t)$ are
continuous, and that the interval $t_3 t_5$ can be subdivided
into one or more parts on each of which $x(t)$ and $y(t)$
have continuous derivatives such that $x'^2 + y'^2 \neq 0$. The
curve represented is then continuous and has a continuously turning tangent except possibly at a finite number
of corners. A much larger variety of curves can be
represented by such parametric equations than by an
equation of the form $y = y(x)$ because the parametric
representation lays no restriction upon the slope of the
curve or the number of points of the curve which may
lie upon a single ordinate, while for an admissible arc
$y = y(x)$ the slope must always be finite and the number
of points on each ordinate at most one.

The mathematician who first made satisfactory
sufficiency proofs in the calculus of variations was Weierstrass, and the ingenious device which he used in his
proofs is called a field. For the problems which we are
considering in this chapter a field F is a region of the xy-
plane with which there is associated a one-parameter

family of straight-line segments each intersecting once a fixed curve D, and which have the further property that through each point (x, y) of F there passes one and but one of the segments. The curve D may be either inside the field, or outside as illustrated in Figure 5, and as a special case it may degenerate into a single fixed point.

The whole plane is a field when covered by a system of parallel lines, the curve D being in this case any straight line or curve which intersects all of the parallels. The plane with the exception of a single point 0 is a field when covered by the rays through 0, and 0 is a degenerate curve D. The tangents to a circle do not cover a field since through each point outside of the circle there pass two tangents, and through a point inside the circle there is none. If, however, we cut off half of each tangent at its contact point with the circle, leaving only a one-parameter family of half-rays all pointing in the same direction around the circle, then the exterior of the circle is a field simply covered by the family of half-rays.

At every point (x, y) of a field F the straight line of the field has a slope $p(x, y)$, the function so defined being called the slope-function of the field. The integral I^* with this slope-function in place of p in its integrand has a definite value along every arc C_{35} in the field having equations of the form (11), as we have seen on page 25. We can prove with the help of the formulas of the last section that the integral I^* associated in this way with a field has the two following useful properties:

The values of I^* are the same along all curves C_{35} in the field F having the same end-points 3 and 5. Furthermore along each segment of one of the straight lines of the field the value of I^* is equal to the length of the segment.

To prove the first of these statements we may consider the curve C_{35} shown in the field of F of Figure 5. Through every point (x, y) of this curve there passes, by hypothesis, a straight line of the field F intersecting D, and the formula (8) of page 26, applied to the one-parameter family of straight-line segments so determined by the points of C_{35}, gives

$$I^*(C_{35}) = I^*(D_{46}) - I(E_{56}) + I(E_{34}) .$$

The values of the terms on the right are completely determined when the points 3 and 5 in the field are given, and are entirely independent of the form of the curve C_{35} joining these two points.   This shows that the value $I^*(C_{35})$ is the same for all arcs C_{35} in the field joining the same two end-points, as stated in the theorem.

The second property of the theorem follows from the fact that along a straight-line segment of the field the differentials dx and dy satisfy the equation $dy = p\ dx$, and the integrand of I^* reduces therefore to $\sqrt{1+p^2}dx$ which is the integrand of the length integral.

We now have the mechanism necessary for the sufficiency proof which was the objective of this section.   We wish to show that a straight-line segment E_{12} joining a pair of points 1 and 2 is shorter than every other arc joining these points.   For that purpose let us consider the field formed by covering the whole xy-plane by the lines parallel to E_{12}.   When C_{12} is an arc joining 1 with 2 in this field and defined by equations in the parametric form (11) the properties just deduced for the integral I^* give

$$I(E_{12}) = I^*(E_{12}) = I^*(C_{12}) = \int_{s_1}^{s_2} \cos \theta\ ds ,$$

and the difference between the values of I along C_{12} and E_{12} is therefore

$$I(C_{12}) - I(E_{12}) = \int_{s_1}^{s_2} (1 - \cos \theta) \, ds \geqq 0 \;.$$

The equality sign can hold only if C_{12} coincides with E_{12}. For when the integral in the last equation is zero we must have $\cos \theta = 1$ at every point of C_{12}, from which it follows that C_{12} is tangent at every point to a straight line of the field and satisfies the equation $dy = p \, dx$. Such a differential equation can have but one solution through the initial point 1 and that solution is E_{12}. We have proved therefore that the length $I(C_{12})$ of C_{12} is always greater than that of E_{12} unless C_{12} is coincident with E_{12}.

We may emphasize again here that the sufficiency proof just given is considerably more inclusive than that of Section 12, page 23, since it clearly shows that a straight line joining the points 1 and 2 is not only shorter than all other admissible arcs $y = y(x)$ joining these points but also shorter than every other curve with the same end-points defined by equations in the parametric form (11).

14. *The shortest arc joining a point to a curve.* If a fixed point 1 and a fixed curve N are given instead of two fixed points it is clear that the shortest arc joining them must again be a straight-line segment, but this property alone is not sufficient to insure a minimum length. There are two further conditions on the shortest line from a point to a curve for which we shall find very interesting analogues in connection with the problems considered in later chapters.

Let the equations of the curve N in Figure 6 be written in terms of a parameter τ in the form

$$x = x(\tau) , \qquad y = y(\tau) ,$$

and let τ_2 be the parameter value defining the intersection point 2 of N with a shortest arc E_{12} joining 1 with N. The arc E_{12} must evi-

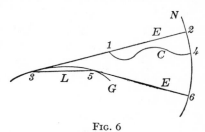

dently be a straight-line segment, since it is certainly a shortest arc joining 1 with 2 if it has this property with respect to curves joining 1 with N. The length of the

FIG. 6

straight-line segment joining the point 1 with an arbitrary point $(x(\tau), y(\tau))$ of N is a function $I(\tau)$ which must have a minimum at the value τ_2 defining the particular line E_{12}. It is clear that the formula (7) of page 25 is applicable to the one-parameter family of straight lines joining 1 with N when in that formula we replace C by the point 1 and D by N. Since along C the differentials dx, dy are then zero it follows that the differential of the function $I(\tau)$ along the arc E_{12} is

$$dI = \frac{dx + p\,dy}{\sqrt{1 + p^2}} \,\bigg|\, 2$$

where the bar indicates that the value of the preceding expression is to be taken at the point 2. Since for a minimum the differential dI must vanish it follows that at the point 2 the differentials dx, dy of N and the slope

p of E_{12} satisfy the condition $dx + p\,dy = 0$, and hence that these two curves must intersect at right angles.

Even a straight-line segment through 1 and intersecting N at right angles may not be a shortest arc joining 1 with N, as may be seen with the help of the familiar string property of the evolute of N. The segments of the straight lines perpendicular to N cut off by N and its evolute G in Figure 6 form a family to which the formula (8) of page 26 is applicable. If in that formula we replace the curve C by G and D by N then

$$I(E_{64}) - I(E_{32}) = I^*(N_{26}) - I^*(G_{35}) .$$

But the integrals in the second member of this formula have the values

$$I^*(N_{26}) = \int_{s_1}^{s_2} \cos \theta \, ds = 0, \quad I^*(G_{35}) = I(G_{35})$$

since $\cos \theta = 0$ along N where the straight lines of the family meet N at right angles, and $\cos \theta = 1$ along the envelope G to which these lines are tangent. Hence from the next to last equation we have the formula

$$I(E_{32}) = I(G_{35}) + I(E_{56}) .$$

This is the string property of the evolute, for it implies that the lengths of the arcs E_{32} and $G_{35} + E_{56}$ are the same, and hence that the free end 6 of a string fastened at 3 and allowed to wrap itself around the evolute G will describe the curve N.

It is evident now that the segment E_{12} cannot be a shortest line from 1 to N if it has on it a contact point 3 with the evolute G of N. For the composite arc $E_{13} + G_{35} + E_{56}$ would in that case have the same length as

E_{12}, and the arc $E_{13}+L_{35}+E_{56}$, formed with the straight line segment L_{35}, would be shorter than E_{12}. It follows then that

If an arc E_{12} intersecting the curve N at the point 2 is to be the shortest joining 1 with N it must be a straight line perpendicular to N at the point 2 and having on it no contact point with the evolute G of N.

We should note the exceptional case when the evolute G has no branch at the point 3 extending toward the point 2. This happens when N is a circle and the evolute G degenerates into a point, or when the evolute has a cusp at 3 with its point directed toward the point 2. In these cases the proof which has just been made does not hold, but it can still be shown that when $I(E_{12})$ is a minimum the point 3 cannot lie between 1 and 2. It can coincide with 1 only in the exceptional case when the envelope G has at 3 no branch extending toward the point 2.

We might continue indefinitely to seek further characteristic properties of the shortest arc joining a point 1 to a curve N, but it will be more satisfactory if we can prove that those already found actually insure a minimum. We infer readily from Figure 6, page 31, that when the end-point 1 lies between 3 and 2 there is adjoining E_{12} a region F of the plane which is simply covered by the normals to N which are near to E_{12}. An analytic proof of this statement is given for a more general case in Section 60, page 156, but for the present we may be content with our inference of it from the figure. The region F so covered by the normals to N forms a field such as was described in Section 13, page 27. The integral I^* formed with the slope function $p(x, y)$ of the field in its integrand is independent of the path and has the same value as I along the straight-line segment E_{12} of the field. It has furthermore the value zero on every arc of N since the straight lines of the field are all perpendicular to N and its integrand therefore vanishes identically along that curve. Hence for an arbitrarily selected arc C_{14} in F joining 1 with N, as shown in Figure 6, page 31, we have

$$I(E_{12}) = I^*(E_{12}) = I^*(C_{14} + N_{42}) = I^*(C_{14}) \ ,$$

and the difference between the lengths of C_{14} and E_{12} is

$$I(C_{14}) - I(E_{12}) = I(C_{14}) - I^*(C_{14}) = \int_{s_1}^{s_2} (1 - \cos \theta) ds \ .$$

We may prove as in the preceding section that this difference is positive unless C_{14} is identical with E_{12}, so that we have established the following theorem:

For a straight-line segment E_{12} perpendicular to the curve N at the point 2 and not touching the evolute G of N there exists a neighborhood F in which E_{12} is shorter than every other arc joining 1 with N.

It is provable, though we shall not undertake it here, that the arc E_{12} still has this minimum property when the point 3 coincides with 1 and the envelope G has no branch extending from 3 toward the point 2, the only exception being the case when N is a circle. If N is a circle and 1 its center then all of the radii are equal in length and it can be shown that they are shorter than the other curves which join 1 to N.

15. *The shortest arc from a point to an ellipse.* An interesting application of the results of the preceding section is afforded by the problem of determining the

shortest curve joining a given point 1 to an ellipse whose equations may be taken in the parametric form

$$x = a \cos t, \qquad y = b \sin t \qquad (0 \le t \le 2\pi) .$$

We know that there must surely be such a shortest curve since the function

$$\phi(t) = \sqrt{(x_1 - a \cos t)^2 + (y_1 - b \sin t)^2} ,$$

representing the distance from the point 1 to a movable point on the ellipse, must have a minimum at some value t_2 on the interval $0 \le t \le 2\pi$, i.e., at some point 2 on the ellipse. The straight line E_{12} joining 1 with 2 is then as short as every other straight line drawn from 1 to the ellipse. If C_{14} is another curve joining 1 to a point 4 on the ellipse we have the relations

$$I(E_{12}) \le I(E_{14}) < I(C_{14})$$

between the lengths of the straight lines E_{12}, E_{14}, and of the curve C_{14}.

In order to characterize the shortest arc E_{12} more explicity let us consider the evolute of the ellipse. The radius of curvature of the ellipse has the value

$$R = \frac{(x'^2 + y'^2)^{\frac{3}{2}}}{x'y'' - x''y'} = \frac{1}{ab}(a^2 \sin^2 t + b^2 \cos^2 t)^{\frac{3}{2}} ,$$

and the equations of the evolute are

$$\xi = x - R \sin \tau = \frac{a^2 - b^2}{a} \cos^3 t ,$$

$$\eta = y + R \cos \tau = \frac{b^2 - a^2}{b} \sin^3 t ,$$

where (ξ, η) is the running point on the evolute, and τ is the angle between the x-axis and the tangent to the ellipse at the point whose parameter value is t. The evolute has the well-known form shown in Figures 7 and 8. It is evident that the number of normals which can be drawn from 1 to the ellipse is the same as the number of tangents from 1 to the evolute.

In case the point 1 is inside the evolute there are four normals through it, as is shown in Figure 7, and of these only the straight line E_{12} and $E_{12'}$ can furnish shortest distances since the others both have points of contact with the evolute between 1 and the ellipse. We know further that each of the arcs E_{12} and $E_{12'}$ does furnish a relative minimum

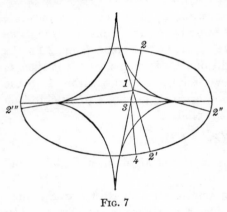

FIG. 7

in the sense that each has a neighborhood F in which at least it is shorter than every other arc joining 1 to the ellipse. The shortest of all the arcs joining 1 to the ellipse, the one furnishing the so-called absolute minimum, must be one of these arcs, and we see readily in Figure 7 that E_{12} is the one.

We may prove the last statement more explicitly by noting in Figure 7 that the normal E_{34} from the intersection point 3 of E_{12} with the x-axis is by symmetry equal in length to E_{32}. The broken line $E_{31}+E_{12'}$ lies within a field F of normals about E_{34} in which the sufficiency proof of the last section applies, and we have therefore the relation

$$I(E_{31}) + I(E_{12'}) > I(E_{34}) = I(E_{31}) + I(E_{12})$$

which shows that the length $I(E_{12'})$ is greater than $I(E_{12})$.

If the point 1 is outside of the evolute there are but two normals from it to the ellipse, as shown in Figure 8. One of these only, the segment E_{12} of the figure, furnishes a relative minimum and it must also in this case provide the absolute minimum.

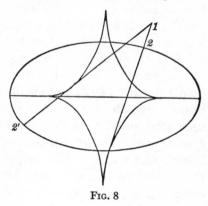

From these results it is clear that on each normal to the ellipse the x-axis marks the point beyond which the segments of the

Fig. 8

normal no longer furnish an absolute minimum. When a point 1 is given the shortest line of all from it to the ellipse is the unique normal segment through 1 which has on it no contact point with the evolute and no intersection with the x-axis. There are a number of exceptional cases when the point 1 is on the x-axis or the evolute which the reader will readily analyze for himself. In particular we may notice that on the normals to the two points at the ends of the major axis of the ellipse the evolute has branches receding from the ellipse. This is an illustration of the exceptional form for the envelope G mentioned in the last section.

16. *The shortest arc joining two curves.* With the results of the preceding sections in mind it is not difficult to solve the problem of determining the shortest distance between two curves. Let M and N be two given curves intersected by a third arc E_{12} in the points 1 and 2, respectively, as shown in Figures 9 and 10. If E_{12} is to furnish a minimum distance from M to N it must be a straight-line perpendicular to them at the intersection points 1 and 2, since it is evident that E_{12} must furnish a minimum among the arcs joining M with 2, or N with 1.

A further condition on the curvatures of M and N at the points 1 and 2 is, however, necessary. Let the points 3 and 4 be the centers of curvature of M and N, respectively, on their common normal E. Then the new necessary condition is that these points with the points 1 and 2 must lie in the circular order 4312 on the straight line E, no coincidences being permitted except possibly that 3 may fall upon 4. By circular order is meant the order of the points on the line E when its two ends at

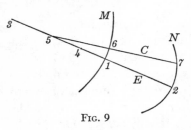

Fig. 9

infinity are thought of as joined together to form a closed arc.

To prove the necessary condition which has just been stated suppose that the point 4 is on the segment 312, as shown in Figure 9, so that the circular order of the four points is not 4312. Let 5 be a point of 312 so selected that the interval 35 contains neither 1 nor 4. Then in every neighborhood F of E_{52} there is an arc C_{57} with length less than that of E_{52}, according to the

theorem of page 33, since the center of curvature 4 of
N lies between 5 and 2.   On the other hand, when the
neighborhood F is sufficiently small, the segment C_{56} of
this arc is longer than E_{51} since 5 lies between 1 and the
center of curvature 3 of M, according to the theorem of
page 34.   From the inequalities

$$I(C_{56})+I(C_{67})<I(E_{51})+I(E_{12}) , \quad I(C_{56})\geqq I(E_{51})$$

expressing these relationships it follows by subtraction
that C_{67} is shorter than E_{12}, and hence that in this case
E_{12} could not furnish a minimum.

When the points 3 and 4 coincide the arc E_{12} may or
may not be the shortest joining M with N.   Only a more
elaborate study than we shall make here could lead to a
conclusion in that case.

Let us assume now
that the four points lie
in the required order
4312 on the line E with
4 distinct from 3, as
shown in Figure 10,
and let us select again

FIG. 10

arbitrarily a point 5 between 4 and 3.   From the theorem
on page 34 we know that in a suitably small neighborhood
F of E_{52} the length of E_{52} is shorter than that of every other
arc joining 5 with N.   Let C_{67} be an arbitrarily selected
arc in F intersecting M and N in 6 and 7, respectively.
We can extend C_{67} to the point 5 by means of an arc C_{56}
not longer than E_{51}, since the center of curvature 3 of M
is between 5 and 1. If C_{67} is not identically E_{12}, then

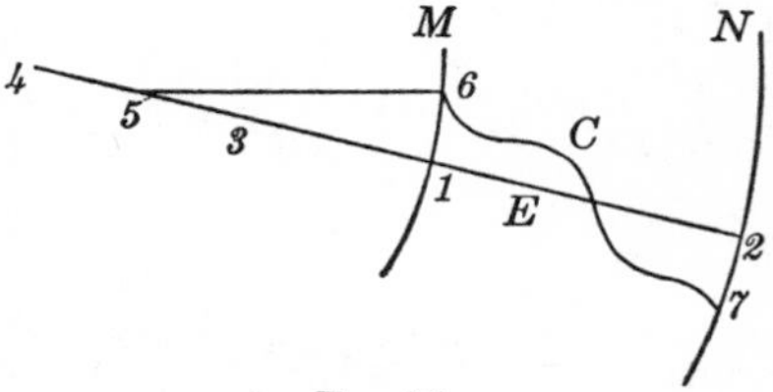

$$I(C_{56})+I(C_{67})>I(E_{51})+I(E_{12}), \ I(C_{56})\leqq I(E_{51}).$$

From these inequalities it follows by subtraction that every arc C_{67} joining M with N in F has length not less than that of E_{12}. Our conclusions are therefore as follows:

The shortest arc joining two curves **M** *and* **N** *must be a straight line* E_{12} *perpendicular to these curves at its endpoints 1 and 2. The centers of curvature 3 and 4 of M and N, respectively, on the line E must lie in the circular order 4312 with 1 and 2, no coincidences being permitted except that 4 may possibly fall upon 3.*

If an arc E_{12} *has these properties, with the additional assumption that 4 is distinct from 3, then there is a neighborhood F of* E_{12} *such that the length of* E_{12} *is surely less than that of every other arc in F joining M with N.*

CHAPTER III

THE BRACHISTOCHRONE PROBLEM

17. *Its significance as an illustration.* The brachisto-chrone problem is historically the most interesting of all the special problems mentioned in Chapter I since as we have there seen it gave the first impetus to systematic research in the calculus of variations. Since the time of the Bernoulli brothers it has been used with great regularity as an illustration by writers on the subject, and it is in many respects a most excellent one. Unfortunately in the forms originally proposed by the Bernoullis it does not require the application of an important necessary condition for a minimum which was first described by Jacobi in 1837, more than a century after the calculus of variations began to be systematically studied.[6] A special case of this condition is the restriction on the position of the center of curvature in the problem of finding the shortest arc from a point to a curve, as described in the theorem on page 33 of the last chapter. It is perhaps at first surprising that the significance of such a simple instance of the condition escaped the early students of the calculus of variations, but a study of the older memoirs soon impresses one with a realization of the serious difficulties encountered with the methods originally used. Throughout the eight-eenth century, investigators in the calculus of variations for the most part desisted when they had found the forms, or in many cases the differential equations only, of the minimizing curves which they were seeking.

It is natural at first sight to suppose that a straight line is the path down which a particle will fall in the shortest time from a given point 1 to a second given point 2, because a straight line is the shortest distance between the two points, but a little contemplation soon convinces one that this is not the case. John Bernoulli explicitly warned his readers against such a supposition when he formally proposed the brachistochrone problem in 1696. The surmise, suggested by Galileo's remarks on the brachistochrone problem, that the curve of quickest descent is an arc of a circle, is a more reasonable one, since there seems intuitively some justification for thinking that steepness and high velocity at the beginning of a fall will conduce to shortness in the time of descent over the whole path. It turns out, however, that this characteristic can also be overdone; the precise degree of steepness required at the start can in fact only be determined by a suitable mathematical investigation.

The first step which will be undertaken in the discussion of the problem in the following pages is the proof that a brachistochrone curve joining two given points must be a cycloid. We are familiar with the cycloid as the arched locus of a point on the rim of a wheel which rolls on a horizontal line, as shown in Figure 11. It turns out that

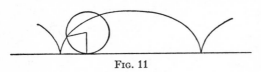

FIG. 11

the brachistochrone must consist of a portion of one of the arches turned upside down, and the line on the under side of which the circle rolls must be located at just the proper height above the given initial point of fall.

When these facts have been established we are at once faced by the problem of determining whether or not such a cycloid exists joining two arbitrarily given points. Fortunately a modification by Schwarz (1845–1923) of a method due to the Bernoulli brothers will enable us to prove that two points can always be joined by one and but one cycloid of the type desired.

When these results had been attained the eighteenth-century student was content with his progress, but we cannot be so easily satisfied because we know that in other problems of the calculus of variations further conditions on the minimizing arc are required which are quite different in character from those which have so far been described. Our doubts for this particular problem will be removed, however, by a so-called sufficiency proof which will definitely establish the fact that the time of descent from a given point 1 to a given point 2 on a suitably chosen cycloid is shorter than that on every other curve joining those two points. The method used is again that of Weierstrass, a special case of which we have already considered in Section 13 on page 27 of the last chapter. The argument there given and the one which we shall see in the case of the brachistochrone are excellent illustrations of the type of proof which is effective for more general problems of the calculus of variations.

18. *The analytic formulation of the problem.* In order to discuss intelligently the problem of the brachistochrone we should first deduce for ourselves the integral which represents the time required by a particle to fall under the action of gravity down an arbitrarily chosen curve joining two fixed points 1 and 2. It is agreed that the initial velocity v_1 at the point 1 is given

in advance, and that the particle is to fall without friction on the curve and without resistance in the surrounding medium. If the effects of friction or a resisting medium are to be taken into account the brachistochrone problem becomes a much more complicated one.

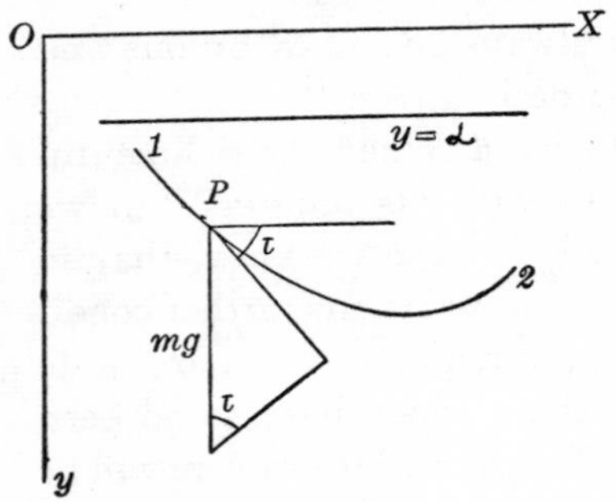

FIG. 12

Let m be the mass of the moving particle P in Figure 12 and s the distance through which it has fallen from the point 1 along the curve of descent C in the time t. In order to make our analysis more convenient we may take the positive y-axis vertically downward, as shown in the figure. The vertical force of gravity acting upon P is the product of the mass m by the gravitational acceleration g, and the only force acting upon P in the direction of the tangent line to the curve is the projection $mg \sin \tau$ of this vertical gravitational force upon that line. But the force along the tangent may also be computed as the product md^2s/dt^2 of the mass of the particle by its acceleration along the curve. Equating these two values we find the equation

$$\frac{d^2s}{dt^2} = g \sin \tau = g\frac{dy}{ds}$$

in which a common factor m has been discarded and use has been made of the well-known calculus formula $\sin \tau = dy/ds$.

To integrate this equation we follow a customary procedure and multiply each side by $2\,ds/dt$. The antiderivatives of the two sides are then easily found, and since they can differ only by a constant we have

$$(1) \qquad \left(\frac{ds}{dt}\right)^2 = 2gy + c \ .$$

The value of the constant c can be determined if we remember that the values of y and $v = ds/dt$ at the initial point 1 of the fall are y_1 and v_1, respectively, so that for $t = 0$ the last equation gives

$$v_1^2 = 2gy_1 + c \ .$$

With the help of the value of c from this equation, and the notation

$$(2) \qquad a = y_1 - \frac{v_1^2}{2g} \ ,$$

equation (1) becomes

$$(3) \qquad \left(\frac{ds}{dt}\right)^2 = 2gy - 2gy_1 + v_1^2 = 2g(y - a).$$

An integration now gives the following result which we have been seeking:

The time T required by a particle starting with the initial velocity v_1 to fall from a point 1 to a point 2 along a curve is given by the integrals

$$(4) \qquad T = \frac{1}{\sqrt{2g}} \int_0^l \frac{ds}{\sqrt{y - a}} = \frac{1}{\sqrt{2g}} \int_{x_1}^{x_2} \sqrt{\frac{1 + y'^2}{y - a}}\,dx$$

where l is the length of the curve and $a = y_1 - v_1^2/2g$.

It is clear that an arc which minimizes one of the integrals (4) expressing T will also minimize that integral when the factor $1/\sqrt{2g}$ is omitted, and vice versa. **Let us** therefore use the notations

$$(5) \qquad I = \int_{x_1}^{x_2} f(y, y')dx, \qquad f(y, y') = \sqrt{\frac{1+y'^2}{y-a}}$$

for our integral which we seek to minimize and its integrand. Since the value of the function $f(y, y')$ is infinite when $y = a$ and imaginary when $y < a$ we must confine our curves to the portion of the plane which lies below the line $y = a$ in Figure 12. This is not really a restriction of the problem since the equation $v^2 = (ds/dt)^2 = 2g(y-a)$ deduced above shows that a particle started on a curve with the velocity v_1 at the point 1 will always come to rest if it reaches the altitude $y = a$ on the curve, and it can never rise above that altitude. For the present we shall restrict our curves to lie in the half-plane $y > a$. In a later section of this chapter we shall see what happens when curves are permitted which have points in common with the line $y = a$.

In our study of the shortest-distance problems in the last chapter the arcs to be considered were taken in the form $y = y(x)$ $(x_1 \leqq x \leqq x_2)$ with $y(x)$ continuous on the interval $x_1 \leqq x \leqq x_2$, and the interval could furthermore be subdivided into parts on each of which the derivative $y'(x)$ is continuous. *An admissible arc* for the brachistochrone problem will always be understood to have these properties besides the additional one that it lies entirely in the half-plane $y > a$. For an admissible function, however, we retain always the definition given on page 18 of the preceding chapter. Our problem is then to find among the admissible arcs joining the points 1 and 2, one which minimizes the integral I.

19. *A first necessary condition.* As in our study of the problems of determining shortest distances let us suppose that a particular admissible arc E_{12} with the equation

$$y = y(x) \qquad (x_1 \leqq x \leqq x_2)$$

actually furnishes a minimum for the integral I, and let us then seek to determine its properties. If $\eta(x)$ is an admissible function having $\eta(x_1) = \eta(x_2) = 0$ then the family of arcs

$$(6) \qquad y = y(x) + a\eta(x) \qquad (x_1 \leqq x \leqq x_2)$$

contains E_{12} for the parameter value $a = 0$, and for small values of a consists entirely of admissible arcs passing through the points 1 and 2. We must not let a be too large, as otherwise the corresponding curve of the family might lie partly above the line $y = a$.

Among the values

$$(7) \qquad I(a) = \int_{x_1}^{x_2} f(y + a\eta,\ y' + a\eta')dx$$

of the integral I along the arcs of the family (6) the particular value $I(0)$, which is the value along E_{12}, must be a minimum, and we have therefore the necessary condition $I'(0) = 0$. The value of the derivative $I'(0)$ found by differentiating equation (7) with respect to a and then setting $a = 0$, is

$$(8) \qquad I'(0) = \int_{x_1}^{x_2} \{f_y\eta + f_{y'}\eta'\}dx \ ,$$

where f_y and $f_{y'}$ are the partial derivatives of $f(y, y')$ with the arguments y, y' belonging to the minimizing arc E_{12}. If we make use of the easily derived formula

$$\eta f_y = \frac{d}{dx}(\eta \int_{x_1}^{x} f_y\, dx) - \eta' \int_{x_1}^{x} f_y\, dx$$

and the fact that $\eta(x_1) = \eta(x_2) = 0$, the expression (8) takes the form

$$I'(0) = \int_{x_1}^{x_2} \left\{ f_{y'} - \int_{x_1}^{x} f_y \, dx \right\} \eta' \, dx \ .$$

This must vanish for every family of the form (6), i.e., for every admissible function $\eta(x)$ having $\eta(x_1) = \eta(x_2) = 0$, and we find ourselves again in a position to make use of the fundamental lemma of page 20. From that lemma it follows that

For every minimizing arc E_{12} there must exist a constant c such that the equation

$$(9) \qquad f_{y'} = \int_{x_1}^{x} f_y \, dx + c$$

holds at every point of E_{12}. On each sub-arc of E_{12} where the tangent turns continuously we must have

$$(10) \qquad \frac{d}{dx} f_{y'} = f_y \ .$$

The last equation may be readily deduced from equation (9) by differentiation, since on a sub-arc of E_{12} where the tangent turns continuously the function f_y is continuous and the integral in (9) has the derivative f_y.

The equation (10) is the famous differential equation deduced by Euler in 1744 and called after him Euler's differential equation.[7] Its solutions have been named extremals because they are the only curves which can give the integral I a maximum or a minimum, i.e., an extreme value. We shall in the following pages apply the term extremal only to those solutions $y(x)$ which have continuously turning tangents and continuous second derivatives $y''(x)$.

We have not so far made any assumption concerning the existence of a second derivative $y''(x)$ along our minimizing arc. When there is one, however, we can carry out the differentiations indicated in equation (10) and obtain

$$\frac{d}{dx}f_{y'}-f_y=f_{y'y}y'+f_{y'y'}y''-f_y=0$$

from which it follows that along a minimizing arc with a second derivative $y''(x)$ we have

$$\frac{d}{dx}(f-y'f_{y'})=y'(f_y-f_{y'y}y'-f_{y'y'}y'')=0$$

and hence also

(11) $$f-y'f_{y'}=\text{constant.}$$

The reasoning by means of which equation (11) has been derived is valid not only for the particular integrand function $f(y, y')=(1+y'^2)^{\frac{1}{2}}/(y-a)^{\frac{1}{2}}$ of the brachistochrone problem but also for an arbitrary function $f(y, y')$ of the two variables y and y' which with its derivatives has suitable continuity properties. One can further verify readily that the proofs of equations (9) and (10) hold without alteration not only for this case but also for the still more general integral I to be studied in Chapter V for which the integrand is assumed to be a function $f(x, y, y')$ of the three variables x, y, y'. It is evident, therefore, that the results of this section have great generality and that they may be applied to a wide variety of problems in the calculus of variations. One should note that equation (11) cannot be expected to hold when x occurs in the integrand function, since in the differentiations made to obtain it the function f was supposed to contain only the variables y and y'.

20. *Application to the brachistochrone problem.* For the special case of the brachistochrone problem the integrand $f(y, y')$ and its partial derivatives are readily found to be

$$(12) \quad f = \sqrt{\frac{1+y'^2}{y-a}} \,,\; f_y = -\tfrac{1}{2}\sqrt{\frac{1+y'^2}{(y-a)^3}} \,,\; f_{y'} = \frac{y'}{\sqrt{(y-a)(1+y'^2)}}$$

and the condition (9) of the preceding section, which must hold along every minimizing arc E_{12}, can therefore be expressed in the form

$$\frac{y'}{\sqrt{1+y'^2}} = \sqrt{y-a}\left\{ c - \tfrac{1}{2}\int_{x_1}^{x}\sqrt{\frac{1+y'^2}{(y-a)^3}}\,dx \right\}.$$

The second member of this equation, which we may denote by $\phi(x)$, is a continuous function of x since both $(y-a)$ and the integral in it are continuous along E_{12}. By solving the equation for y' we find that

$$y' = \frac{\phi}{\sqrt{1-\phi^2}}$$

is also continuous. Turning again to the next to last equation with this result we see now that the second member $\phi(x)$ of that equation has also a continuous derivative, since when y is continuous and has a continuous derivative the same is true of both $(y-a)$ and the integral occurring in it. Hence y' in the last equation must have a continuous derivative and we find the following result:

For the brachistochrone problem a minimizing arc E_{12} lying entirely below the line $y = a$ can have no corners and *must have continuous curvature. Analytically stated, this means that the derivatives $y'(x)$, $y''(x)$ exist and are continuous along E_{12}.*

Since we now know that the second derivative $y''(x)$ exists along the minimizing arc we can be sure that the equation $f - y'f_{y'} = $ constant deduced in the preceding section also holds along it. When the values of f and its derivative $f_{y'}$ for the brachistochrone problem are substituted from (12) this equation becomes

$$(13) \qquad f - y'f_{y'} = \frac{1}{\sqrt{(y-a)(1+y'^2)}} = \frac{1}{\sqrt{2b}} ,$$

the value of the constant being chosen for convenience in the form $1/\sqrt{2b}$.

The curves which satisfy the differential equation (13) may be found in the customary manner by solving the equation for $y' = dy/dx$ and separating the variables, but we shall find them more easily if we profit by the experience of others and introduce a new variable u defined by the equation

$$(14) \qquad y' = -\tan \frac{u}{2} = -\frac{\sin u}{1+\cos u} .$$

From the differential equation (13) it follows then, with the help of some simple trigonometry, that along a minimizing arc E_{12} we must have

$$y - a = \frac{2b}{1+y'^2} = 2b \cos^2 \frac{u}{2} = b(1+\cos u) ,$$

$$\frac{dx}{du} = \frac{dx}{dy}\frac{dy}{du} = 2b \cos^2 \frac{u}{2} = b(1+\cos u) ,$$

$$x = a + b(u + \sin u) ,$$

where the last equation is found from the next to last
by an integration and a is the new constant so introduced.
It will be shown in the next section that curves which
satisfy the first and third of these equations are the
cycloids described in the following theorem:

*A curve down which a particle, started with the initial
velocity v_1 at the point 1, will fall in the shortest time to a
second point 2 is necessarily an arc having equations of the
form*

(15) $$x - a = b(u + \sin u), \qquad y - a = b(1 + \cos u).$$

*These represent the locus of a point fixed on the circum-
ference of a circle of radius b as the circle rolls on the lower
side of the line $y = a = y_1 - v_1^2/2g$. Such a curve is called
a cycloid.*

21. *Cycloids.* The fact that the equations (15)
represent a cycloid of the kind described in the theorem
is easily proved. For let a circle of radius b begin to
roll on the line $y = a$ at
the point whose co-ordi-
nates are (a, a), as shown
in Figure 13. After a
turn through an angle of
u radians the point of
tangency is at a distance

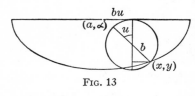

Fig. 13

bu from (a, a), and the point which was the lowest in
the circle has rotated to the point (x, y). The values
of x and y may now be readily calculated in terms of u
from the figure, and they are found to be those given by
equations (15).

The fact that the curve of quickest descent must be a
cycloid is the famous result discovered by James and

John Bernoulli in 1697 and announced at approximately
the same time by a number of other mathematicians.
The cycloid and its remarkable properties had been the
subject of much study in the seventeenth century.
Huygens (1629–95) in
particular had shown
that the evolute of a
cycloid CC' shown in
Figure 14 is a second
cycloid of exactly the
same size situated in the
position which is dotted
in the figure, and further
that the time of descent

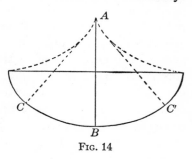

Fig. 14

of a particle starting at rest at the point C and fall-
ing along the cycloid to the lowest point B is the same
no matter what the position of the starting-point C on
the cycloid arc may be. We know from the string
property of an evolute, proved on page 32 of the last
chapter, that if a pendulum of length $4b$ is so suspended
at A that the string is compelled to wrap itself around
the dotted evolute, then the pendulum bob will oscillate
on an arc CC' of the cycloid. From the isochronous
property of Huygens it follows that the period of oscillation will be the same no matter how great or how small
the amplitude of the oscillation from C to C' may be.
This has been considered a very important discovery
for the clockmaker, though one rarely finds apparatus
on his mantel which is built upon this principle.

These and other properties of the cycloid were well
known before the end of the seventeenth century. That
the cycloid should also be the solution of the brachisto-

chrone problem was regarded with wonder and admiration by the Bernoulli brothers. Somewhat freely translated the comment of John was "With justice we admire Huygens because he first discovered that a heavy particle falls on a cycloid in the same time always, no matter what the starting-point may be. But you will be petrified with astonishment when I say that exactly this same cycloid, the tautochrone of Huygens, is the brachistochrone which we are seeking." And with rhetorically sustained enthusiasm James remarks "Thus for this curve, which has been investigated by so many mathematicians that apparently nothing further concerning it could remain to be discovered, we find a new property, as it were an indication of its desire that no obligation might be incurred to future centuries but that it might attain the pinnacle of perfection at the end of the present one at whose beginning its birth was celebrated and among whose researches there have fallen to its lot the discovery of all of its mensurable properties and many other beautiful characteristics."

At the present time our mathematics is usually less emotionally expressed, but in thought at least we can share the pleasure of the brothers in their interesting discovery. It is clear when we think intuitively about the brachistochrone, that the straight line could hardly have been expected to be the curve of quickest descent because it is evident that a particle should fall more rapidly on a curve which begins more steeply and imparts a higher velocity at the beginning of the fall. But just what degree of initial steepness is desirable in order to avoid too gradual a slope near the end of the fall could hardly have been foreseen.

22. *The unique cycloid through two points.* We have seen in the preceding section that the minimizing arc which we are seeking must be one of the cycloids generated by rolling a circle on the lower side of the line $y = a$, but we have so far had no assurance that it is always possible to find such an arc joining two arbitrarily selected points 1 and 2. Unless we can prove that these points can be so joined we cannot be sure that there is such a thing as a curve of quickest descent. The question is not one which can be neglected, as will appear more clearly in the problem of the determination of a surface of revolution of minimum area to be studied in the next chapter. For that problem two points 1 and 2 can easily be selected in such relative positions that there is no minimizing arc joining them and expressible in the form $y = y(x)$. For the brachistochrone problem it happens fortunately, however, that we can establish the following theorem:

Through every pair of points 1 and 2 below the line $y = a$, and not in the same vertical, there passes one and but one cycloid generated by a circle rolling on the lower side of that line.

We may always interchange the numbering of the points 1 and 2 if necessary so that we have $x_2 > x_1$. When $x_2 = x_1$ so that the points 1 and 2 are in the same vertical line the curve of quickest descent from one to the other is the straight-line segment joining them, as we shall see in a later section.

Analytically the problem before us is the determination of four values a, b, u_1, u_2, satisfying the four equations

(16)
$$x_1 - a = b(u_1 + \sin u_1), \qquad x_2 - a = b(u_2 + \sin u_2),$$
$$y_1 - a = b(1 + \cos u_1), \qquad y_2 - a = b(1 + \cos u_2),$$

The first two of these impose the condition that the cycloid shall pass through the point 1 for the parameter value u_1, and the last two have a similar meaning for u_2 and the point 2. The problem of showing that these equations have solutions a, b, u_1, u_2 has been successfully discussed analytically by several writers.[8] We shall here convince ourselves that there is one and but one set of solutions (a, b, u_1, u_2) of the equations (16), or in other words that there is a unique cycloid of the family (15) passing through the points 1 and 2, by a geometrical argument due to Schwarz which is an extension of one given for the special case when $v_1 = 0$ by the brothers Bernoulli themselves.

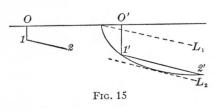

FIG. 15

In the first place let us draw an arbitrary one of the cycloids (15) and intersect it by a line $1'2'$ parallel to the straight line joining 1 with 2, as shown in Figure 15. If we move the line $1'2'$, keeping it always parallel to 1 2, from the position L_1 to the tangent position L_2, the ratio of the segment $0'1'$ to the segment $1'2'$ increases from zero to infinity and passes once only through the value of the corresponding ratio of 0 1 to 1 2. In the position which gives the equality of these ratios the lengths of $0'1'$ and $1'2'$ are not necessarily equal respectively to those of 0 1 and 1 2. By changing the value of b, however, the cycloid (15) can be expanded or contracted into another similar to itself and having the same center, and the new segments $0''1''$ and $1''2''$ corresponding to $0'1'$ and $1'2'$ will have the same ratio as before. If the value of b is properly chosen the segments $0''1''$ and $1''2''$ will be exactly equal to 0 1 and 1 2. Changing the value of a merely slides the cycloid from right to left, or vice versa, along the line $y = a$, so that by making a suitable selection of a the points of $1''$ and $2''$ can be made to coincide with 1 and 2, and the validity of our theorem is established.

23. *The construction of a field.* In their study of the problems of the calculus of variations the mathematicians of the eighteenth and early nineteenth centuries did not distinguish clearly between sets of conditions which are necessary for a minimum, and those which are sufficient to actually insure a minimum. The result was that for a long period after the discovery of each new necessary condition on the minimizing arc the students of the subject were satisfied that they had complete solutions of their problems. Weierstrass was the first to point out the desirability of a sufficiency proof. In the decade preceding 1879, during which he lectured frequently on the calculus of variations at the University of Berlin, he discovered a new necessary condition, and he was able to prove that this condition with the three previously known were sufficient for a minimum. It is true that slight alterations must be made in the necessary conditions, like the one which we have seen on page 4 for the minimum of a function $f(x)$, in order that they may also be sufficient.

For the particular case of the brachistochrone problem we have seen that the minimizing arc must be a cycloid, and that there is a unique cycloid joining the points 1 and 2, but we do not know as yet that this cycloid will always be a curve of quickest descent. In the problem of finding the shortest distance from a fixed point 1 to a curve, studied on page 30 and following pages, the solution was necessarily a straight line normal to the curve, but it was also necessary that the point 1 should not lie farther away from the curve than the center of curvature on this normal. May it not be, in the case of the brachisto-chrone, that on the cycloid arc joining 1 with 2 there is a critical point 3 associated with 1 beyond which the minimizing property of the arc no longer holds? The answer in this case is no, though we shall see in Chapters IV and V that in general such a point is to be expected· To make sure of our answer a sufficiency proof is neces-sary, and in this section we are to discuss one of the important steps in the proof, the construction of what is called a field of extremals.

Let us consider the particular extremal arc E_{12} of our problem joining the points 1 and 2. It is necessarily one of the cycloids of the family (15) characterized by two special values a_0, b_0 of the constants a, b. If we keep $a = a_0$ fixed and let b vary the resulting one-parameter family of inverted cycloid arches

$$(17) \qquad x - a_0 = b(u + \sin u) , \qquad y - a = b(1 + \cos u) ,$$
$$(-\pi \leqq u \leqq \pi; \; 0 < b < \infty)$$

will have a common center (a_0, a). On two different cycloids of the family corresponding to the parameter values b_0 and b the same value of the variable u defines two points (x_0, y_0) and (x, y) whose co-ordinates satisfy the equations

$$\frac{x - a_0}{x_0 - a_0} = \frac{b}{b_0} , \qquad \frac{y - a}{y_0 - a} = \frac{b}{b_0} .$$

These equations mean geometrically that on all the radii through the center (a_0, a) the two cycloids cut off segments which are proportional to b and b_0. The cycloids are therefore similar figures, the one corresponding to the

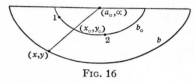

FIG. 16

value b being formed from the arc E_{12} corresponding to b_0 by stretching it along the radii through the center (a_0, a) into a new

curve whose radii are b/b_0 times as long as those of E_{12}.

A region of the plane which is simply covered by a one-parameter family of extremals is called a *field of extremals* or simply a *field*. The argument just made shows that for our brachistochrone problem the half-plane $y > a$ is such a field, simply covered by the one-parameter family of concentric cycloids (17). Analytically this is equivalent to saying that for each point (x, y) in the field the equations (17) are satisfied by one and but one pair of values b, u which we may denote by $b(x, y)$, $u(x, y)$. The slope at the point (x, y) of the extremal of the field through that point is found from equation (14) to have the value

(18) $$p(x, y) = -\tan \frac{u(x, y)}{2}$$

and the function $p(x, y)$ so defined is called the *slope function* of the field.

In order to carry through the operations of the succeeding sections we should convince ourselves that the functions $b(x, y)$, $u(x, y)$, $p(x, y)$ described above as belonging to the field are continuous and have continuous derivatives. The proofs that they have such properties are given in the next section which the reader may omit if he is willing to take results of this sort for granted.

24. *Properties of the field functions.* If we can prove that the function $u(x, y)$ of the last section is continuous and has continuous derivatives then the same will be true of the functions

$$p(x, y) = -\tan \frac{u(x, y)}{2}, \quad b(x, y) = \frac{y-a}{1+\cos u}$$

obtained from (18) and the second of the equations (17). The denominator in the expression for $b(x, y)$ is different from zero since on the cycloid arches of the field we have always $-\pi < u < +\pi$. Let us therefore concentrate our attention upon $u(x, y)$.

For the determination of $u(x, y)$ the equations (17) give by division the relation

$$(19) \qquad \frac{x-a_0}{y-a} = \frac{u+\sin u}{1+\cos u}$$

which has the form $g(x, y) = h(u)$. If we change x and y to $x+\Delta x$ and $y+\Delta y$ and denote by Δu the change thus caused in u we find readily

$$g(x+\Delta x, y+\Delta y) - g(x, y) = h(u+\Delta u) - h(u)$$

which with the help of Taylor's formula becomes

$$(20) \qquad g_x\Delta x + g_y\Delta y = h_u\Delta u$$

where the derivatives g_x, g_y have the arguments $x+\theta\Delta x$, $y+\theta\Delta y$, and h_u the argument $u+\theta\Delta u$, with $0<\theta<1$, as provided by Taylor's formula. The derivative h_u is readily found to have the value

$$h_u = \frac{2+2\cos u + u\sin u}{(1+\cos u)^2} = \frac{1+\frac{u}{2}\tan\frac{u}{2}}{\cos^2\frac{u}{2}}$$

which is not less than unity on the cycloids of the field since on them we have always $-(\pi/2) < u/2 < \pi/2$. Hence we may solve equation (20) for Δu and find

$$(21) \qquad \Delta u = \frac{g_x}{h_u}\,\Delta x + \frac{g_y}{h_u}\,\Delta y$$

which shows that Δu approaches zero with Δx and Δy and hence that the function $u(x, y)$ is continuous.

To find the partial derivative u_x we set $\Delta y = 0$ in equation (21) and evaluate the limit of the quotient $\Delta u/\Delta x$ as Δx approaches zero. A similar process gives u_y, and the results are easily seen to be

$$u_x = \frac{g_x(x, y)}{h_u(u)}\,, \qquad u_y = \frac{g_y(x, y)}{h_u(u)}\,.$$

These functions are continuous in x and y since $u(x, y)$ is continuous.

The functions $u(x, y)$ and $b(x, y)$ are also continuous at each point (x, y) on the boundary $y = a$ of the field distinct from (a_0, a). In order to prove this we note first that the values (u, b) involved in our discussion all satisfy the inequalities $-\pi \le u \le \pi$, $0 < b < \infty$ indicated just after the equations (17). Consider then the neighborhood of points (u, b) satisfying the inequality

$$\sqrt{(u - u_0)^2 + (b - b_0)^2} < \epsilon\,,$$

where ϵ is arbitrary and $(u_0, b_0) = (\pm\pi, b_0)$ is a pair whose corresponding point (x_0, y_0) by means of equation (17) is on the line $y = a$; and let δ be the minimum distance from the point (x_0, y_0) to the points (x, y) defined by values (u, b) outside of this neighborhood. The points (x, y) satisfying the inequality

$$\sqrt{(x - x_0)^2 + (y - y_0)^2} < \delta$$

and lying on or below the line $y = a$ all have values $u(x, y)$, $b(x, y)$ corresponding to them which lie in the ϵ-neighborhood described above, since no (u, b) outside this neighborhood can correspond to a point (x, y) whose distance from (x_0, y_0) is less than δ. It follows that $u(x, y)$ and $b(x, y)$ are continuous at (x_0, y_0) since, as just shown, we may make (u, b) lie in arbitrarily small ϵ-neighborhoods of (u_0, b_0) by taking (x, y) sufficiently near to (x_0, y_0).

A similar argument shows that $b(x, y)$ has the limit zero at $(x, y) = (a_0, a)$; and when (x, y) approaches this center along a curve one may prove by means of equation (19) that the limit of $u(x, y)$ is the unique solution of the equation

$$\frac{dx}{dy} = \frac{u + \sin u}{1 + \cos u} ,$$

where dx/dy is the reciprocal of the slope of the curve of approach at the point (a_0, a).

25. *Two important auxiliary formulas.* In order to discuss further the minimizing properties of our cycloids we shall need two formulas which are analogous to the formulas (7) and (8) which were developed for straight-line segments on pages 25 and 26 of the last chapter. If a segment E_{34} of a cycloid varies so that its end-points describe two curves C and D, as shown in Figure 17, then it is possible to find a formula for the differential of the value of the integral I taken along the moving segment, and a formula expressing the difference of the values of I at two positions of the segment.

The equations

$$(22) \quad x = a(t) + b(t)(u + \sin u) , \qquad y = a + b(t)(1 + \cos u)$$

$$(u_3(t) \leqq u \leqq u_4(t))$$

define a one-parameter family of cycloid segments E_{34} when a, b, u_3, u_4 are functions of a parameter t as indicated in the equations. If t varies, the end-points 3 and 4 of this segment describe the two curves C and D whose equations in parametric form with t as independent variable are found by substituting $u_3(t)$ and $u_4(t)$, respectively, in the equations (22). These curves

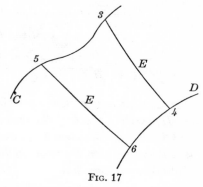

FIG. 17

and two of the cycloid segments joining them are shown in Figure 17.

If we notice that along each of our cycloids the relations

(23)
$$p = \frac{-\sin u}{1+\cos u}, \qquad 1+p^2 = \frac{2}{1+\cos u},$$

$$y - a = b(1+\cos u), \qquad \frac{dx}{du} = b(1+\cos u)$$

hold, then the value of the integral I taken along the particular arc E_{34} is readily found to be

(24)
$$I(E_{34}) = \int_{u_3}^{u_4} \sqrt{\frac{1+p^2}{y-a}} \frac{dx}{du} \, du = u\sqrt{2b} \; \Big|_3^4,$$

and its differential is evidently expressible in terms of db, du_3, du_4. We could calculate da and these three differentials from the four equations, two for the curve C and two for D, deduced from (22) by substituting the two functions $u_3(t)$ and $u_4(t)$ and differentiating. It is easier, however, to note that when u is a function of t, as well as a and b, the equations (22) and (23) give

$$dx = da + (u + \sin u)db + b(1 + \cos u)du ,$$

$$dy = \qquad (1 + \cos u)db - b \sin u \, du ,$$

$$\frac{dx + p \, dy}{\sqrt{y-a} \sqrt{1+p^2}} = \frac{da + u \, db + 2b \, du}{\sqrt{2b}} = \frac{da}{\sqrt{2b}} + d(u\sqrt{2b}) .$$

Hence we have from (24) the following important result:

If a cycloid segment E_{34} varies so that its end-points 3 and 4 describe simultaneously two curves C and D, as shown in Figure 17, then the value of the integral I taken along E_{34} has the differential

$$(25) \qquad dI = d(u\sqrt{2b}) \Big|_3^4 = \frac{dx + p \, dy}{\sqrt{y-a} \sqrt{1+p^2}} \Big|_3^4 .$$

At the points 3 and 4 the differentials dx, dy in this expression are those belonging to C and D, while p is the slope of the segment E_{34}.

If the symbol I^* is now used to denote the integral

$$(26) \qquad I^* = \int \frac{dx + p \, dy}{\sqrt{y-a} \sqrt{1+p^2}}$$

then by an integration of the formula (25) with respect to t from t_3 to t_5 we find the further result that

The difference between the values of I at two different positions E_{34} and E_{56} of the variable cycloid segment, shown in Figure 17, is given by the formula

$$(27) \qquad I(E_{56}) - I(E_{34}) = I^*(D_{46}) - I^*(C_{35}) .$$

The formulas (25) and (27) are the analogues for cycloids of the formulas (7) and (8) of pages 25 and 26 for straight lines. We shall see that they have numerous applications in the theory of brachistochrone curves.

26. *The invariant integral of the field.* Consider now the field F of concentric inverted cycloid arches simply covering the half-plane below the line $y = a$, as described in Section 23, page 57, and let us substitute the slope-function $p(x, y)$ of the field in place of p in the integrand of the integral I^* defined in equation (26). Along each arc C_{35} in the field, of the type described on page 27 and having equations

$$(28) \qquad x = x(t) , \qquad y = y(t) \qquad (t_3 \leqq t \leqq t_5) ,$$

the integral I^* so formed has a perfectly definite value found by substituting the values of x, y, dx, dy from these equations in its integrand and integrating with respect to t from t_3 to t_5. Through each point of the arc C_{35} there passes a unique cycloid arc of the field inter- secting also the verti- cal line D through the common central point (a_0, a) of all the cycloids of the field, as shown in Figure 18. The two

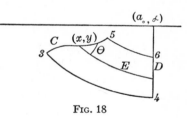

FIG. 18

points where the particular cycloids through 3 and 5 in- tersect D may be designated by 4 and 6. The equation (27) is then applicable, and it shows that the value $I^*(C_{35})$ must depend only upon the end-points 3 and 5, and not at all upon the form of the arc C_{35} joining them, since the other three terms in the equation have this property. It should also be noticed that along a cycloid arc of the field the differentials dx and dy satisfy the equation $dy = p\, dx$ and for such values the integrands of I^* and I

are easily seen to be identical. Hence we have the following theorem:

The integral I^* formed with the slope-function $p(x, y)$ of the field F in its integrand takes the same value along all arcs in the field joining the same two end-points 3 and 5. Furthermore along a cycloid arc E_{12} of the field the value of I^* is the same as that of the integral I.

If we make use of the formulas (9) of page 26 we find for the integral I^* the very compact and useful expression

$$(29) \qquad I^* = \int \frac{\cos \theta \, ds}{\sqrt{y-a}} \, .$$

The angle θ at a point (x, y) of an arc C_{35} of the field is now the one between the tangents to C_{35} and to the extremal of the field through (x, y), while s is the length of arc measured along C_{35}.

27. *The sufficiency proof.* With the help of the properties of the integral I^* described in the preceding section it is now very easy to prove that the time required by a particle to fall down a cycloid arc from a point 1 to a point 2 is shorter than that required on every other arc of the type (28) described on page 65 joining those two points and lying below the line $y = a$. The proof to be presented is in essence the one originally given by Weierstrass in his lectures. In outward form it is somewhat different because use is made of the invariant integral I^* whose importance and convenience in this connection were first emphasized by Hilbert some thirty years after Weierstrass had made his original demonstration.

Let E_{12} be the unique cycloid arc joining the points 1 and 2 whose existence was discussed in Section 22, page 55. According to the results of Section 23, page 57, it is one of the arcs of a field F of concentric inverted cycloid arches simply covering the half-plane $y > a$. Let C_{12} be an arbitrarily selected arc in F, also joining the points 1 and 2, and having equations of the parametric form (28). Since the integral I^* formed with the slope-function $p(x, y)$ of the field has the same value as I along the cycloid arc E_{12} of the field, and since it has further this same value on all arcs whatsoever in F joining 1 with 2, it follows that

$$I(E_{12}) = I^*(E_{12}) = I^*(C_{12}) .$$

Hence with the help of the expressions (5) and (29) for I and I^* the difference between the values of I on the arcs C_{12} and E_{12} is seen to be

$$(30) \quad I(C_{12}) - I(E_{12}) = I(C_{12}) - I^*(C_{12}) = \int_{s_1}^{s_2} \frac{1 - \cos \theta}{\sqrt{y - a}} ds .$$

This difference is evidently always positive or zero so that we surely have $I(C_{12}) \geqq I(E_{12})$. The equality sign can hold only if $\cos \theta = 1$ along the arc C_{12}, i.e., only if C_{12} is at every point tangent to a cycloid of the field and consequently satisfies everywhere the equation $dy = p(x, y)dx$. In that case, however, C_{12} must coincide with E_{12}, since the differential equation of the first order $dy = p(x, y)dx$ has one and but one solution through the point 1, and that is E_{12} itself.

In connection with these results it should be noted that we started out on page 46 to consider only admissible arcs of the form $y = y(x)$. In making the sufficiency proof of the present section it has been quite as easy, however, to allow the comparison arcs C_{12} to be expressed in the parametric form (28) which permits them to wander in less restricted fashion from the point 1 to the point 2. Our conclusions may therefore be summarized in the following manner:

Let an initial velocity $v_1 \neq 0$ and two fixed points 1 and 2 not on the same vertical line be given. For a particle starting at the point 1 with the initial velocity v_1 the curve of quickest descent from the point 1 to the point 2 must be an arc of a cycloid generated by a point fixed on the circumference of a circle rolling on the lower side of the line $y = a$, where $a = y_1 - v_1^2/2g$. There is one and but one such cycloid, E_{12}, joining 1 with 2, and the time of descent on E_{12} is shorter than the time on every other arc of the type (28) joining 1 with 2 and lying below the line $y = a$.

The sufficiency theorem for curves in the parametric form gives a much stronger conclusion than was demanded in the original statement of the problem.

28. *The case when the initial velocity is zero.* We have seen on page 46 that the integrand of the integral I becomes infinite at a point for which $y = a$, and for that reason we have agreed hitherto to consider only curves which lie everywhere below the line $y = a$. In so doing we have unfortunately excluded the interesting case when the particle starts from rest, since for $v_1 = 0$ the value of $a = y_1 - v_1^2/2g$ is y_1 and the initial point 1 itself lies on the line $y = a$. This case may also be treated, however, with the help of results already attained and a limiting process which seems to have been first used by Weierstrass.

Suppose now that $v_1 = 0$ and let E_{12} be again the unique cycloid arc joining the point 1 on the line $y = a$ with the point 2, as shown in Figure 19. Through each point 3 of an arbitrarily selected curve C_{12} in the half-plane $y \geqq a$

there passes a unique cycloid E_{34} of the field F intersecting the vertical line D through the center (a_0, a) in a point 4.   The sum $I(C_{13}) + I(E_{34})$ varies continuously as the point 3 moves from 1 to 2 along the arc C_{12}, beginning with the value $I(E_{12}) + I(E_{27})$ and ending at the value $I(C_{12}) + I(E_{27})$. Hence if we can show that this sum does not decrease then $I(C_{12}) \geqq I(E_{12})$.

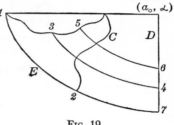

FIG. 19

To show that it is increasing consider a second point 5 near to 3 on the arc C_{12}.   By means of the formula (27) of page 64 applied to the last two terms of the expression

$$[I(C_{15}) + I(E_{56})] - [I(C_{13}) + I(E_{34})] = I(C_{35}) + I(E_{56}) - I(E_{34})$$

we see that this difference has the value

$$I(C_{35}) + I^*(D_{46}) - I^*(C_{35}) \ .$$

Since the vertical line D_{46} is orthogonal to the cycloids of the field the equation $dx + p \ dy = 0$ holds along it and the formula for I^* on page 64 shows that this integral vanishes along D_{46}.   The remaining difference $I(C_{35}) - I^*(C_{35})$ is positive or zero by the same argument which showed this to be true for the difference (30), and zero only when the arc C_{35} is coincident with a cycloid arc of the field F.   If every partial arc C_{35} of the curve C is coincident with a cycloid of the field, then C_{12} must coincide with E_{12}, and we see therefore that $I(C_{12}) > I(E_{12})$ unless C_{12} and E_{12} are identical.

The argument of the last few paragraphs can also be used in the case when $v_1 \neq 0$ if we wish to consider comparison arcs which have points in common with the line $y = a$.[37] It is furthermore applicable when the points 1 and 2 are in the same vertical line E_{12}. In the latter case every point 3 of an arbitrary curve C_{12} may be joined to 2 by a cycloid or vertical line E_{32}. The sum $I(C_{13}) + I(E_{32})$ increases from the value $I(E_{12})$ to the value $I(C_{12})$ as the point 3 traverses the arc C, so that the vertical line E_{12} is in this case the curve of quickest descent.

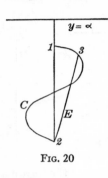

Fig. 20

29. *The path of quickest descent from a point to a curve. First necessary conditions.* At the conclusion of his now famous solution of the brachistochrone problem, published in 1697, James Bernoulli proposed to other mathematicians, but to his brother in particular, a number of further questions. One of them was the problem of determining the arc down which a particle, starting with a given initial velocity, will fall in the shortest time from a fixed point to a fixed vertical straight line. This is a special case of the more general problem of determining the brachistochrone arc joining a fixed point 1 to an arbitrarily chosen fixed curve N. The material deduced in the preceding sections, especially the formulas (25) and (27) of page 64, enable us to find readily a solution of this problem.

Let the point 1, the curve N, and the path E_{12} of quickest descent be those shown in Figure 21, and let the given initial velocity at the point 1 again be v_1. It is clear that the minimizing arc E_{12} must be one of the cycloids already discussed in preceding pages, since if it is the path of quick-est descent from 1 to the curve N it must also be the path of quickest descent from 1 to the intersection point 2. Thus we have at once a first necessary condition on our minimizing arc,

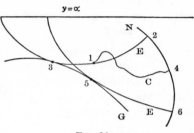

FIG. 21

a condition which must, however, be supplemented in this case by two further properties of a different character.

Since there is a unique cycloid generated by a circle rolling on the lower side of the line $y = a$ and joining 1 with a second point below that line, it follows that there is a one-parameter family of such cycloid arcs joining the point 1 to the different points of the curve N. The values of the integral I taken along the different members of the family must have a minimum along E_{12} if that arc is to be the curve of quickest descent, and the differential of I must vanish along E_{12}. To calculate this differential we apply the formula (25) of page 64 to the one-parameter family of cycloids joining 1 with N, replacing the curve C of that formula by the fixed point 1 and the curve D by N. We find then

$$dI = \frac{dx_2 + p_2\, dy_2}{\sqrt{y_2 - a}\,\sqrt{1 + p_2^2}}$$

for the value of the differential along E_{12}, since at the point 1 the differentials dx_1 and dy_1 are both zero. Since the slopes p_2 and dy_2/dx_2 of the curves E_{12} and N at their intersection point 2 must make this expression vanish, we may summarize the conditions so far found for our minimizing arc as follows:

For a particle starting at the point 1 with the initial velocity v_1 the path of quickest descent from 1 to a curve N is necessarily an arc E_{12} of a cycloid generated by a point fixed on the circumference of a circle rolling on the lower side of the line $y = y_1 - v_1^2/2g$. The path E_{12} must furthermore be cut at right angles by the curve N at their intersection point 2.

30. *The envelope theorem and the analogue of Jacobi's condition.* For the cycloids of the brachistochrone problem which we are studying there is a theorem which corresponds in a very interesting way to the well-known string property of the evolute of a curve described on page 32 of the last chapter. To deduce this property we make use of the fact, which will be proved in Section 33 for a more general case, that the curve N has adjoining E_{12} a one-parameter family of our cycloid arcs each of which, like E_{12}, is cut at right angles by the curve N. This family may have an envelope G as shown in Figure 21. If in the formula (27) of page 64 the arc C_{35} is replaced by G_{35}, and D_{46} by N_{26}, then the difference between the values of I along the two cycloids E_{56}, E_{32} of the family is seen to be

$$I(E_{56}) - I(E_{32}) = I^*(N_{26}) - I^*(G_{35}) .$$

Along the arc N_{26} the integral I^* has the value zero since the direction of N is at each point perpendicular to the intersecting cycloid and hence the numerator $dx + p \, dy$ of the integrand of I^* vanishes identically along N. On the other hand the arc G_{35} is at each point tangent to the corresponding cycloid, so that $dy = p\ dx$ along G_{35} and the value $I^*(G_{35})$ is equal to $I(G_{35})$, as we readily see from the formula (26) for I^* on page 64. The last equation is therefore equivalent to

$$(31) \qquad I(E_{32}) = I(G_{35}) + I(E_{56}).$$

This is the very remarkable analogue of the string property of the evolute mentioned above, and like that property it is a special case of a more general theorem associated with the integrals which we shall study in Chapter V. Although a number of special cases of this theorem have long been known, the proofs of it for more general cases were first made by Darboux (1842–1917) and Zermelo in 1894, and by Kneser in 1898.[9] For the special problem of this chapter we may state this theorem as follows:

THE ENVELOPE THEOREM. *Let G be the envelope of a one-parameter family of cycloids E each of which is cut at right angles by a curve N, as shown in Figure 21 on page 71, the cycloids all being generated by circles rolling on the lower side of the line $y = a$. Then the time required by a particle to fall along the arc E_{32} in the figure is the same as that required on the composite arc $G_{35} + E_{56}$ provided that the initial velocity at the point 3 is in each case $v_3 = \sqrt{2g(y_3 - a)}$.*

This result is the interpretation in other words of the equation (31) above. The initial velocity at the point 3 must be as specified in the theorem since the integrals (4) on page 45 always express the time of fall of a particle whose velocity at the altitude y is $v = \sqrt{2g(y - a)}$, as indicated by equation (3).

The envelope theorem as stated above enables us to prove without further difficulty a third necessary condition which must be satisfied by the arc of quickest descent from 1 to N. In Figure 21 the point 3 is shown outside of the arc E_{12}. If, however, it lay between 1 and 2 then $I(E_{12})$ could not be a minimum, since in every neighborhood of E_{12} there would be a composite arc $E_{13} + G_{35} + E_{56}$ down which the time of fall would be the same as that for E_{12}, and we could always replace G_{35} by an arc V_{35} down which the time of fall would be shorter since G_{35} is not one of our cycloid arcs, as it must be if it is to furnish a minimum time of fall between its end-points. To make quite sure of the statement that G_{35} is not one of the cycloid arcs we note that the equations

$$p = -\tan \frac{u}{2} , \quad x = a + b(u + \sin u) , \quad y = a + b(1 + \cos u)$$

have only one solution u, a, b when x, y, p are given, and therefore that there is one and but one cycloid through a given point (x, y) and having there the direction p. The arc G_{35} is tangent at each point to a cycloid and hence cannot be one of them.

A further necessary condition on the curve E_{12} of quickest descent from a point 1 to a curve N is that the arc E_{12} shall not have on it a contact point with the envelope G, shown in Figure 21 on page 71, of the one-parameter family of cycloid arcs which are cut at right angles by the curve N.

This necessary condition which has just been deduced is analogous to the condition mentioned in the first paragraph of this chapter as having been discovered for the general theory by Jacobi. The orthogonality of the arc N to the cycloid E_{12}, and the necessary condition on the position of the contact point 3 of E_{12} with the envelope G, correspond in a striking way to the similar conditions described on page 33, for the problem of finding the shortest distance from a point to a curve. An exceptional case should be noted here as in the problem of shortest distances. If the envelope G has no branch extending from the point 3 toward the curve N then the proof of Jacobi's condition cannot be made as above by means of the envelope theorem. Other methods can, however, be used to show that the value $I(E_{12})$ can never be a minimum when the contact point 3 of E_{12} with the envelope G lies between 1 and 2. It will in general be a minimum when the point 3 is at 1 and the envelope has no branch projecting toward the point 2. That this exceptional case can happen quite frequently is evident if we take for the curve N an orthogonal trajectory to the cycloids through the point 1. The envelope G is then the point 1 itself. It is also provable that we can always expect the envelope G to have a cusp at the point 3 with branches projecting away from the curve N when the radius of curvature of N has a minimum at the point 2.

31. *Sufficient conditions.* The one-parameter family of cycloids orthogonal to the curve N does not simply cover the whole half-plane below the line $y = a$ in the sense that through each point of the half-plane there passes one and but one cycloid of the family, as one readily sees by an examination of the region near the envelope G in Figure 21, page 71. It is clear intuitively from the figure, however, that if the point 3 lies exterior to the arc E_{12}, as indicated in the figure, then there is a neighborhood of this arc through each point of which

passes a unique cycloid of the family. The neighborhood so covered is a field F in the sense described on page 59. We need ultimately to have a more rigorous proof of the existence of this field, but we may for the present be content with the assurance given by our interpretation of the figure. In Section 60, page 156, there is an analytical proof of the existence of a field for a more general case which completely establishes the existence of the field here needed.

Let C_{14} in Figure 21 of page 71 be an arc in the field F joining the point 1 to the curve N. The sufficiency proof of Section 27, page 66, when modified slightly enables us to prove that the value $I(C_{14})$ is greater than $I(E_{12})$ except when C_{14} coincides with E_{12}. The integral I^* formed with the slope-function $p(x, y)$ of the field in its integrand has as before the same value as I along the arc E_{12}, and has also this same value along all other curves in the field with the same end-points 1 and 2. Consequently

$$I(E_{12}) = I^*(E_{12}) = I^*(C_{14}) + I^*(N_{42}) .$$

Along the arc N_{42} the cycloids of the field are perpendicular to N so that $dx + p(x, y)dy = 0$ and $I^*(N_{42})$ vanishes. We have therefore, as before,

$$I(C_{14}) - I(E_{12}) = I(C_{14}) - I^*(C_{14}) = \int_{s_1}^{s_4} \frac{1 - \cos \theta}{\sqrt{y - a}} ds .$$

This expression is always positive unless C_{14} coincides with E_{12}, as was shown on page 67. Our result is then as follows:

Let an initial velocity $v_1 \neq 0$, a fixed point 1, and a fixed curve N be given. For a particle starting at the point 1 *with the initial velocity v_1 an arc E_{12} of quickest descent from 1 to the curve N, such as is shown in Figure 21, page 71, must have the following properties:*

1. It must be an arc of a cycloid generated by a point fixed on the circumference of a circle rolling on the lower side of the line $y = y_1 - v_1^2/2g$.

2. It must cut N at right angles at their intersection point 2.

3. It must not contain a contact point 3 with an envelope G of a one-parameter family of such cycloids orthogonal to the curve N of which it is a member.

For an arc E_{12} with these properties there is a neighborhood F such that the time of descent on E_{12} is shorter than that for every other arc in F joining 1 with N.

It is understood that the statements in this theorem and their proofs must be modified somewhat in case the envelope G of the family of cycloids orthogonal to N has no branch at 3 projecting toward the curve N. It is provable that in that case the point of contact 3 may coincide with 1 and $I(E_{12})$ still be a minimum.

If the curve N is a vertical straight line, as in the query proposed by James Bernoulli mentioned on page 70, then the family of cycloids orthogonal to N is a concentric family with no envelope G, such as is shown in Figure 19 of page 69, and the field F is the whole half-plane $y > a$. There is always in this case an arc E_{12} satisfying the three requirements of the theorem and it furnishes a minimum with respect to all the curves below the line $y = a$ joining 1 with N. It is clear that the Jacobi condition 3 of the theorem is now always satisfied since the family of concentric cycloids has no envelope.

32. *The path of quickest descent from a curve to a point.*
It is natural to expect that the problem of determining
the curve of quickest descent from a fixed curve M to
a fixed point 2 will be similar to that of the preceding
sections where the particle started at a fixed point and
fell to a fixed curve.   The criteria encountered, however,
are quite different, the difference being due to the fact
that the line $y = y_1 - v_1^2/2g$ on which the cycloid-
generating circles roll is no longer fixed when the co-
ordinate y_1 of the point 1 is variable on a curve. Lagrange
himself at first thought that the solution of the problem
should be a cycloid orthogonal to the curve M, probably
because he did not specify any too clearly the initial
velocity v_1 which the particle should have as it leaves the
curve M.[10] His reasoning was criticized by Borda
(1733–99) who gave the correct condition on the direction
of the curve M at its intersection with the cycloid,[11] and

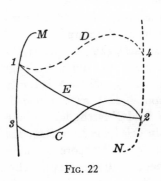

FIG. 22

Lagrange thereupon showed
how his own analysis could be
modified to justify the same
result.[12]

We can reduce the new
problem at once to the preceding case by a simple geometrical transformation which
shows clearly the reasons
why the old criteria no longer
apply. Let $v_1 \neq 0$ be the given
velocity with which the par-
ticle is to start from the curve M shown in Figure 22
and let E_{12} be a particular path from M to the point 2.
The time of fall on every other such path C_{32} is the same

as that on a path D_{14} obtained by moving C_{32} parallel to itself until its initial point 3 falls upon 1. If we rotate M through an angle of 180° on the center of the straight line joining 1 with 2 it will arrive at the position N in the figure. Every arc C_{32} joining M with 2 determines in this way an arc D_{14} joining 1 with N, and conversely, and the times of fall on corresponding arcs are the same provided that the particle starts in each case with the initial velocity v_1. Clearly the problem of this section is equivalent to that of determining the path of quickest descent from the fixed point 1 to the fixed curve N.

We know already the characteristic properties of the solution of the latter problem. A minimizing arc, say E_{12}, must be a cycloid generated by a circle rolling on the line $y = y_1 - v_1^2/2g$, and the curve N must cut E_{12} orthogonally at the point 2. This means that the original curve M at its intersection point 1 with E_{12} must have a direction perpendicular to the tangent to E_{12} at 2. The transformation which we have made shows clearly why the direction of M at 1 is determined by the direction of E_{12} at the point 2 instead of at the point 1.

A further necessary condition on the arc E_{12}, corresponding to the condition 3 of the theorem of page 77, is also readily deducible in terms of the curve N. The point of contact of the arc E_{12}, or its extension, with the envelope G of the one-parameter family of cycloids orthogonal to N must actually lie outside this arc if E_{12} is to furnish a path of quickest descent from M to 2.

33. *The determination of the focal point.* The point 3 where the cycloid E_{12} in Figure 21, page 71, touches the envelope G of the family of cycloids orthogonal to the curve N is called the focal point of N on E on account of

its analogy with the focus of a lens or of a curved mirror. The part of the theorem of page 77, which is concerned with the position of this focal point will not be of much practical value unless we have an explicit method of some sort for determining the position of the point on E. In the following paragraphs a geometrical construction for the focal point is described which can be applied to a large class of problems of which the brachistochrone problem is a special case. It is a slight modification of an interesting generalization by Dr. I. A. Barnett[13] of a construction which was made by Professor Mary E. Sinclair[14] for the problem of the surface of revolution of minimum area.

To deduce this criterion we may take the x-axis in the line $y = a$ so that the equations of the cycloids which we wish to consider are

$$(32) \qquad x = a + b(u + \sin u), \qquad y = b(1 + \cos u).$$

If the first of these equations were solved for u as a function of the quotient $(x-a)/b$ and the result substituted in the second, a single equation for the cycloids would be found of the form

$$(33) \qquad y = b \, \phi \left(\frac{x-a}{b} \right).$$

When the function ϕ is left unspecified this equation is a more general one than that of the cycloids, but it turns out nevertheless that for all problems of the calculus of variations whose extremals are representable by an equation of this form the same geometrical construction for the focal point can be applied.

To find the construction let the equation of the curve N be given in the parametric form

$$x = \xi(\tau) , \qquad y = \eta(\tau) .$$

The equations which express the fact that the curve (33) intersects N at right angles can be taken in the form

$$(34) \qquad \eta = b\,\phi(v), \qquad -\frac{\xi'}{\eta'} = \phi'(v), \qquad v = \frac{\xi - a}{b} ,$$

where a new variable v has been introduced to make the analysis more convenient. The solutions of these equations for v, a, b are three functions $v(\tau)$, $a(\tau)$, $b(\tau)$ such that the one-parameter family of curves

$$(35) \qquad y = b(\tau)\phi\!\left(\frac{x - a(\tau)}{b(\tau)}\right)$$

has each member intersecting the curve N orthogonally. According to a well-known rule of the calculus the envelope G of this family can be found by setting equal to zero the partial derivative with respect to τ of the second member of the last equation. If in making this differentiation we replace always ϕ and ϕ' by their values from the relations $y = b\,\phi$, $y' = \phi'$ which hold along the curve (35) we find the equation

$$(36) \qquad a' + \frac{b'}{b}\left(x - a - \frac{y}{y'}\right) = 0$$

for the determination of the focal point, where a' and b' are the derivatives of a and b with respect to τ, and y' the derivative of y with respect to x.

The derivatives a' and b' in the last equation are to be obtained by differentiating the equations (34) with respect to τ and solving the resulting equations with respect to v', a', b'. If the intersection point of the curve (35) with N is again denoted by 2 we may replace everywhere $\phi(v)$, $\phi'(v)$, $\phi''(v)$ by their values y_2/b, y_2', by_2'', in terms of the ordinate y of the curve (35) and its derivatives at that intersection point. We may suppose $\eta(\tau)$ increasing with τ along the curve N so that from the second of equations (34) we have the relations

$$\sqrt{\xi'^2+\eta'^2}=\eta'\sqrt{1+[\phi'(v)]^2}=\eta'\sqrt{1+y_2'^2}$$

when the signs of the radicals are all positive. We also recall that the radii of curvature ρ and r of N and the curve (35), respectively, have the values

$$\rho=\frac{(\xi'^2+\eta'^2)^{\frac{3}{2}}}{\xi'\eta''-\xi''\eta'}, \qquad r=\frac{(1+y'^2)^{\frac{3}{2}}}{y''}.$$

With these remarks in mind we may find without serious difficulty from equations (34) the derivatives

(37)
$$v'=\frac{\eta'r_2}{b\rho_2}, \quad \frac{b'}{b}=\frac{\eta'}{y_2}\left(1-y_2'\frac{r_2}{\rho_2}\right),$$

$$a'=\frac{\eta'}{y_2}\left[\left(x_2-a-\frac{y_2}{y_2'}\right)y_2'\frac{r_2}{\rho_2}-(x_2-a+y_2y_2')\right].$$

The equation (36) for the determination of the focal point (x, y) now becomes

$$\frac{x-\dfrac{y}{y'}-x_2+\dfrac{y_2}{y_2'}}{x_2+y_2y_2'-x+\dfrac{y}{y'}}=-\frac{\rho_2}{y_2'r_2}=-\frac{\rho_2\cos\theta_2}{r_2\sin\theta_2}$$

where $y_2'=\tan\theta_2$ is the slope of the curve (35) at the point 2. With the help of the well-known formulas

$$t=x-\frac{y}{y'}, \quad n=x+y\,y'$$

for the abscissas t and n of the intersections with the x-axis of the tangent and normal at a point (x, y) of a curve, we find finally the equation

(38)
$$\frac{t_2 - t}{t - n_2} = -\frac{\rho_2 \cos \theta_2}{r_2 \sin \theta_2}$$

which is only another form of equation (36).

The last equation justifies the following geometrical construction of the focal point, illustrated in Figure 23.[38]

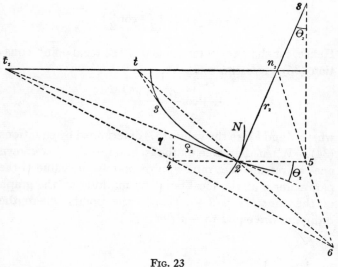

Fig. 23

We draw first the radii of curvature r_2, ρ_2 of the cycloid and the curve N at their intersection point 2, the former of which is negative when calculated from the equations of the cycloid since the derivative y_2'' is negative. Their projections, $-r_2 \sin \theta_2$ and $\rho_2 \cos \theta_2$, on a parallel to the

x-axis through the point 2 are the lines 42 and 25 in the figure. The lines $t_2 4$ and $n_2 5$ meet in a point 6, and the line 62 then cuts the horizontal axis in a point t such that the segments $t_2 - t$ and $t - n_2$ are by similar triangles proportional to 42 and 25. Hence according to equation (38) the contact point of the tangent from t to the cycloid is the focal point 3 of the curve N on the cycloid E.

The parameter value u of this focal point of N on the cycloid (32) is determined analytically by a simple equation. We have, on the cycloid,

$$x - a - \frac{y}{y'} = 2b \left(\frac{u}{2} + \cot \frac{u}{2} \right).$$

Hence, for the parameter value u of the focal point, equation (36) gives the condition

$$\frac{u_3}{2} + \cot \frac{u_3}{2} = -\frac{a'}{2b'}$$

where a' and b' are the values (37) determined by equations (34). When the value of the fraction $-a'/2b'$ is known one can readily determine approximately the value of the parameter u at the focal point by marking on the graph of the function $u/2 + \cot(u/2)$ the points where the ordinates are equal to $-a'/2b'$.

CHAPTER IV

SURFACES OF REVOLUTION OF MINIMUM AREA

34. *Preliminary remarks concerning the problem.* The problem of determining a surface of revolution of minimum area, like the brachistochrone problem, was one of the earliest considered by students of the calculus of variations, and it is one of those which have been most thoroughly studied. It is in many respects the most satisfactory illustration which we have of the principles of the general theory of the calculus of variations in the plane. In spite of the fact that it was proposed early in the eighteenth century and has been restudied at frequent intervals since that time, one finds nevertheless that new results of interest and importance concerning it have been found in very recent years.

In the following pages it will be shown first of all that a plane curve $y = y(x)$ which joins two given points and generates by rotation about the x-axis a surface of revolution of minimum area, must be an arc of a catenary with an equation of the form

(1) $$y = \frac{b}{2}\left[e^{\frac{x-a}{b}} + e^{-\frac{x-a}{b}} \right].$$

The shape of one of these catenaries is shown in Figure 24, but they vary considerably in appearance. One can realize these differ-

Fig. 24

ences experimentally by hanging a chain on two pegs. When the pegs are near together the chain hangs in a catenary which is narrow and vertical, and when the pegs are far apart the curve is broad and flat.   The curve in the figure is of an intermediate type.

If it is admitted that when two points 1 and 2 in the plane are given, a minimizing curve of the form $y = y(x)$ joining them must be one of the catenaries (1), then it devolves upon us at once to find out if the two points can be joined by such a curve.   The analytical discussion of this problem involves computations which will be indicated in a later section, but if we are willing to forego proofs for the moment it will be easy to describe the results geometrically.   In the two-parameter family of catenaries (1) those which pass through the point 1

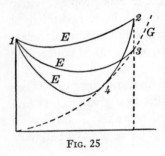

FIG. 25

form a one-parameter family of curves such that one of them passes through 1 in each direction, and this one-parameter family has an envelope G as shown in Figure 25.   It will be proved, as may also easily be inferred intuitively from the figure, that when the initial slope of the catenary at 1 is made to increase from minus infinity to plus infinity the intersection of the catenary with the ordinate through 2 first descends from infinity to the point 3 and then ascends to infinity again.   When the intersection reaches a point 2 above the envelope on its downward journey it belongs to a catenary arc which touches G between 1 and 2, and when it reaches 2 going upward it belongs to a catenary having no such contact point.   We see readily then that every point 2 above the envelope G is joined to 1 by two catenaries of the family (1) only one of which touches the envelope G; a point 2 on G has only one catenary joining it to 1; and a point 2 below G has none. This is a situation very different from the corresponding one for the brachistochrone problem, where there was always a unique cycloid joining two given points, and it is one of the reasons why the catenary problem is so much more typical of the results which may in general be expected for problems of the calculus of variations in the plane.

The point 3 where one of the catenaries E touches the envelope G, in Figure 25, is called the point conjugate to 1 on E. We shall see in a later section that if 4 is a second such contact point on the envelope G, as shown in the figure, then the surface of revolution generated by the composite arc $E_{14}+G_{43}$ is always equal to that generated by E_{13}. This is a very interesting analogue of the string property of the evolute of a curve, and is another instance of the envelope theorem which was justified by Darboux, Zermelo, and Kneser. By means of it we shall be able to prove Jacobi's necessary condition which says that a catenary arc E_{142} having on it a point 4 conjugate to 1 can never furnish a surface of revolution of minimum area. As we have hitherto seen, the problems of the two preceding chapters required no application of Jacobi's necessary condition for the case when the two end-points of the minimizing curve were given in advance. It was for this case, however, that Jacobi originally stated his remarkable condition which distinguishes the calculus of variations in such a striking way from the ordinary theory of maxima and minima of functions of one or more variables.

After seeing that one of the catenaries joining 1 with 2 can never be a solution of our problem, it becomes a matter of some importance to be able to prove that the remaining one, E_{12}, does furnish a minimum of some sort. This will be done by the method of Weierstrass which has already been explained for the problems of the preceding chapter.   There is always a neighborhood of E_{12} in which all other arcs joining 1 with 2 generate larger surfaces of revolution than that generated by E_{12} itself.

The results which have so far been described evidently leave us in some doubt as to what happens when the point 2 lies on or below the envelope G. When 2 is on G, Jacobi's condition says that the unique catenary arc joining 1 with 2 cannot possibly furnish a minimum surface of revolution, and when 2 is below G there is no catenary of the family (1) whatsoever joining 1 with 2. When 2 is on or below G there is in fact no curve represented by an equation of the form $y = y(x)$ which generates a minimum surface of revolution. We shall see that the minimum surface is in this case furnished by the broken line consisting of the two ordinates of the points 1 and 2 and the segment of the x-axis between them.   It is called the Goldschmidt discontinuous solution of the problem, after the man who first discovered it in 1831,[15] discontinuous because its tangent turns discontinuously at its two corners on the x-axis.

It will be proved that the Goldschmidt discontinuous solution, like the minimizing catenary described above, always generates a smaller surface of revolution than those generated by other curves joining its end-points and lying in a sufficiently small neighborhood of it. When the point 2 is above the envelope G both the catenary solution and the discontinuous solution are present, and Dr. H. F. MacNeish showed in 1905[16] how one may distinguish the one of them which generates the smaller surface. Professor Mary E. Sinclair[17] has proved that the smaller one is also smaller than the surfaces of revolution generated by all other curves of a very general type joining the points 1 and 2. The methods of proving these statements in the following pages are somewhat different from the ones used by these writers, but the results established are identical with theirs.

35. *The proof that the minimizing arc is a catenary.* As we have already noted in Chapter I, the integral which we shall have to minimize for the problem of determining a surface of revolution of minimum area is

$$I = \int_{x_1}^{x_2} f(y, y') dx$$

where $f(y, y')$ has now the value

(2) $$f(y, y') = y\sqrt{1 + y'^2}.$$

The curves to be studied for this problem must all lie in the upper half-plane $y \geq 0$ since on an arc with portions below the x-axis the value of the integral I is the difference of the areas generated by the segments above and the segments below the axis, while we are wishing to consider the sum of those areas. If an arc has seg-

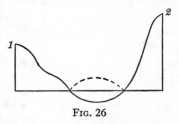

Fig. 26

ments below the x-axis it may always be replaced by one above the axis which will generate the same surface, as is clear from Figure 26. Besides the restriction $y \geq 0$ upon our curves $y = y(x)$ we shall, as in the two preceding examples, assume that all the curves $y = y(x)$ considered are continuous and have tangents which turn continuously except possibly at a finite number of corners. Let us call curves of this sort in the upper half-plane *admissible curves*.

Our problem is then to determine among all admissible arcs joining two given points 1 and 2 one which minimizes the integral I.

The necessary conditions deduced in Section 19, page 47, for a minimizing arc of an integral with an integrand of the form $f(y, y')$, apply without alteration to our present problem. The minimizing arcs must be solutions of the equation

$$f_{y'} = \int_{x_1}^{x} f_y \, dx + c$$

which for the special function (2) takes the form

$$\frac{yy'}{\sqrt{1+y'^2}} = \int_{x_1}^{x} \sqrt{1+y'^2} \, dx + c = s + c \, ,$$

where s is the length of the minimizing arc measured from 1 to the point whose abscissa is x. At a point of the arc where $y > 0$ this equation can be solved for y', giving

$$(3) \qquad\qquad y' = \frac{s+c}{\sqrt{y^2 - (s+c)^2}}$$

and we see at once that at such a point y' is continuous since y and s both have this property. But if y' is continuous then y and s both have continuous derivatives and the equation (3) shows again that y' must also have a continuous derivative. At all points above the x-axis

our minimizing arc has therefore continuous curvature
and no corners.

If we know that along a minimizing arc there is a
continuous derivative y'' then as on page 49 Euler's
equation has the consequence $f - y' f_{y'} =$ constant, which
for the special function (2) takes the form

$$(4) \qquad \frac{y}{\sqrt{1 + y'^2}} = b .$$

By solving this equation for y' and integrating we see
that the solutions of Euler's equation also satisfy the
two equations

$$\frac{dy}{\sqrt{\left(\frac{y}{b}\right)^2 - 1}} = dx, \qquad b \log \left(\frac{y}{b} + \sqrt{\left(\frac{y}{b}\right)^2 - 1} \right) = x - a ,$$

and it follows readily by solving the last one for y that
the extremals for our problem are the catenaries

$$(5) \qquad y = \frac{b}{2}\left[e^{\frac{x-a}{b}} + e^{-\frac{x-a}{b}} \right] = b \, ch \, \frac{x-a}{b} .$$

We use here and in the following pages the customary
symbols $ch\,u$, $sh\,u$ for the hyperbolic cosine and the hyperbolic sine of u defined by the equations

$$ch \, u = \frac{e^u + e^{-u}}{2} , \qquad sh \, u = \frac{e^u - e^{-u}}{2} .$$

No elaborate properties of these functions will be needed,
but it will be helpful later if we notice, while these formulas are before us, that each of them has the other as its
derivative.

When we know that the extremals are the catenaries
of Figure 24, page 85, we see at once that a minimizing

arc $y=y(x)$ with corners is impossible since the corners would have to be on the x-axis, as has already been indicated, and the parts of the minimizing arc above the axis would have to be segments of catenaries which have no points in common with the axis. We have justified, therefore, the following conclusion:

If 1 and 2 are two points in the half-plane $y>0$ then an admissible arc $y=y(x)$ joining them and generating a surface of revolution of minimum area must be a single arc without corners of one of the catenaries (5).

36. *The one-parameter family of catenaries through a point.* Our next step is to determine the number and the character of the catenaries (5) which pass through the two given points 1 and 2. The plan is to find the equation of the one-parameter family of these catenaries passing through the point 1, and then to determine how many of them pass through the second point 2. The equation

$$y_1 = b \; ch \; \frac{x_1-a}{b}$$

is the condition that the catenary (5) shall pass through the point 1. It is satisfied, as one readily verifies, when a and b are expressed in terms of a new parameter a in the form

(6) $$a = x_1 - y_1 \frac{a}{cha} \; , \qquad b = \frac{y_1}{cha} \; ,$$

and the family of catenaries through the point 1 is therefore

(7) $$y = \frac{y_1}{cha} \; ch\left(a + \frac{x-x_1}{y_1} \; cha\right) = y(x, a) \; ,$$

where $y(x, a)$ is simply a convenient symbol for the more complicated expression preceding it.

In deducing the properties of the one-parameter family of catenaries through the point 1, we shall need the first and second derivatives with respect to x and a of the function $y(x, a)$ defined in equation (7). Derivatives with respect to x will be denoted by primes and with respect to a by subscripts, while the subscript 1 will be used to designate values of x, y, y' at the point 1. If we remember the formulas

$$\frac{d}{du} ch\, u = sh\, u \, , \qquad \frac{d}{du} sh\, u = ch\, u$$

mentioned above, we find readily the values

$$y = \frac{y_1}{cha} ch\left(a + \frac{x - x_1}{y_1} cha\right) = y(x, a)$$

(8)
$$y' = sh\left(a + \frac{x - x_1}{y_1} cha\right) , \qquad y_1' = sha$$

$$y_a = \frac{y' y_1'}{cha}\left(x - \frac{y}{y'} - x_1 + \frac{y_1}{y_1'}\right) ,$$

where in calculating the last derivative hyperbolic sines and cosines have been replaced except in one instance by their values in terms of y, y', y_1, y_1' from the first three equations.

In terms of running co-ordinates (X, Y) the tangents to the catenary at the points (x_1, y_1) and (x, y) have the equations

$$Y - y_1 = y_1'(X - x_1) , \qquad Y - y = y'(X - x) .$$

By eliminating X the co-ordinate Y of their point of intersection is found to be

$$Y = \frac{y' y_1'}{y' - y_1'}\left(x - \frac{y}{y'} - x_1 + \frac{y_1}{y_1'}\right) ,$$

so that the derivative y_a has also the value

$$(9) \qquad y_a = \frac{y' - y_1'}{ch\,a}\, Y .$$

This formula justifies at once a very interesting geometrical construction for the point conjugate to 1 on a catenary, discovered in 1860 by Lindelöf (1827–1908).[18] For at the point of tangency (x, y) of a curve $y = y(x, a)$ with the envelope G of the family the derivative y_a must vanish, according to a well-known criterion of the calculus. The co-ordinate Y must be zero, and the tan-

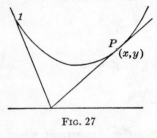

FIG. 27

gents to the catenary at the point 1 and its conjugate (x, y) must therefore meet on the x-axis, as shown in Figure 27. Bolza has proved similarly that the Lindelöf construction for the conjugate point is valid in the more general case when the extremals are represented by an equation $y = b\phi[(x - a)/b]$ such as that discussed in Section 33, page 80.[19]

It is clear from Figure 27 and the formula (9) that as the point (x, y) moves from the point 1 toward the right along the catenary $y = y(x, a)$ the value of y_a is at first positive, and it changes sign only if (x, y) passes a point conjugate to 1. Hence the inequality $y_a > 0$ at a value $x > x_1$ implies that there is no conjugate point between 1 and (x, y) on the catenary, while $y_a < 0$ implies that there is one.

The second derivatives of $y(x, a)$ will be needed only at points (x, y) which are conjugate to 1 and have $x > x_1$.

For such points we see from Figure 27 for the Lindelöf construction that $y_1' < 0$, $y' > 0$. From the formulas (8) we find, then, eliminating always the difference $x - x_1$ by means of the equation $y_a = 0$,

(10)
$$y'' = \frac{y}{y_1^2} ch^2 a > 0 , \qquad y_a' = \frac{y^2 y_1'}{y' y_1^2} ch a < 0 ,$$

$$y_{aa} = \frac{y' y_1'^2}{y_1^2}\left[\left(\frac{y}{y'}\right)^3 - \left(\frac{y_1}{y_1'}\right)^3\right] > 0 .$$

37. *Proofs of the properties of the family.* We are first of all interested to find out what happens to the ordinate of the catenary $y = y(x, a)$ when $x > x_1$ is kept fixed and a varies from $-\infty$ to $+\infty$. If we express this ordinate, from equation (7), in the form

$$y(x, a) = \left(\frac{a}{ch a} + \frac{x - x_1}{y_1}\right) \frac{y_1 \, ch\left[\left(\frac{a}{ch a} + \frac{x - x_1}{y_1}\right) ch a\right]}{\left(\frac{a}{ch a} + \frac{x - x_1}{y_1}\right) ch a} ,$$

we see that it approaches $+\infty$ when a approaches either of the values $\pm \infty$, since from the calculus rule for evaluating an indeterminate form we know that as u approaches plus or minus infinity

$$\lim \frac{u}{ch \, u} = \lim \frac{1}{sh \, u} = 0 .$$

Furthermore the derivative y_a changes from negative to positive whenever it vanishes, on account of the property $y_{aa} > 0$ in formulas (10) of the preceding section, and y_a can therefore vanish only once. We see that when $x > x_1$ is fixed and a varies from $-\infty$ to $+\infty$ the ordinate $y(x, a)$ diminishes from $+\infty$ to a minimum and then increases to $+\infty$ again, since y_a varies from negative values through zero once to positive values.

Let us denote by $g(x)$ the minimum value of the ordinate $y(x, a)$ for a fixed x. The equation $y = g(x)$ defines a curve which we shall presently prove to be the envelope G of the family of catenaries. From the argument of the preceding paragraph it follows that through a point 2 above this envelope there pass two catenaries on which the derivative y_a has opposite signs at 2. According to a remark on page 94 one of these catenaries has a point conjugate to 1 between 1 and 2, the other none. Hence we have the following theorem:

A point 2 above the envelope G in Figure 25 of page 86 is joined to 1 by two catenaries of the family $y = b\ ch[(x-a)/b]$. On one of these there is a point 3 conjugate to 1, on the other none. A point 2 on the envelope G is joined to 1 by a single catenary of the family on which 2 is conjugate to 1. A point 2 below G is joined to 1 by no catenary of the family.

The value of a at which the ordinate $y(x, a)$ has its minimum for a fixed x may be denoted by $a(x)$, and the function so defined satisfies the equation $y_a = 0$. By methods quite similar to those of Section 24, page 60, we may show that $a(x)$ is continuous and has a continuous derivative, so that we have the relations

$$y_a(x,\ a(x)) = 0\ , \qquad y'_a + y_{aa}a' = 0\ ,$$

$$g(x) = y(x,\ a(x))\ , \qquad g'(x) = y' + y_a a' = y' > 0\ ,$$

$$g''(x) = y'' + y'_a a' = \frac{yy_1 y'^3 ch^2 a}{y_1^2 y'^3 - y^2 y_1'^3} > 0\ .$$

The last of these is found by substituting the value of a' from the second and then inserting the values of the second derivatives of y from the equations (10). The fourth equation shows that for $x > x_1$ the curve $y = g(x)$ is everywhere tangent to a catenary and ascending, while the last indicates that its curvature is everywhere positive, as indicated in Figure 25 on page 86.

To justify finally the form of the envelope G shown in the figure we must establish the further properties

$$\lim_{x = x_1} g(x) = 0 , \qquad \lim_{x = x_1} g'(x) = 0 ,$$

$$\lim_{x = +\infty} g(x) = +\infty , \qquad \lim_{x = +\infty} g'(x) = +\infty .$$

To prove those in the first row we notice that as a approaches $-\infty$ the vertex

$$(a, b) = \left(x_1 - y_1 \frac{a}{cha} , \frac{y_1}{cha} \right)$$

of the catenary (7) approaches the point $(x_1, 0)$, and is joined to that point by a line whose slope $-1/a$ approaches zero. Since this vertex is always above the envelope G it follows first that $g(x)$, and then that $g'(x)$, must approach zero as x approaches x_1. The third limit above is evident since the slope $g'(x)$ is positive and increases with x, and $g(x)$ must therefore approach infinity with x. The last one is also true since at the abscissa $x = x_1 - y_1/y_1' + y/y'$ of the point conjugate to 1 on a catenary the slope of the catenary, from equations (8) on page 93, is

$$sh\left(a + \frac{x - x_1}{y_1} cha \right) = sh\left(a - \frac{cha}{sha} + \frac{y}{y'} \frac{cha}{y_1} \right) .$$

This approaches $+\infty$ as a approaches zero through negative values, since $-cha/sha$ approaches $+\infty$. Since the envelope G is tangent to each catenary at the conjugate point it follows that $g'(x)$ also approaches infinity with x.

38. *Two important auxiliary formulas.* In order to discuss further the minimizing properties of our catenaries we might use special methods adapted to catenaries only, as we have done previously for our straight lines and cycloids, but it will be quite as easy to deduce such properties with the aid of two more general formulas which can be applied repeatedly in the study of other special problems of the calculus of variations and also in the more general theory of Chapter V. The special cases of these formulas which we have already seen are the formulas (7) and (8) of pages 25 and 26, and the formulas (25) and (27) of page 64.

For the purpose of developing our new equations let us consider a one-parameter family of extremal arcs

$$(11) \qquad y = y(x, b) \qquad (x_3 \leqq x \leqq x_4)$$

satisfying the Euler differential equation

$$(12) \qquad \frac{\partial}{\partial x} f_{y'} = f_y$$

The partial derivative symbol is now used because there are always the two variables x and b in our equations. If x_3, x_4, and b are all regarded as variables the value of the integral I along an arc of the family is a function of the form

$$I(x_3, x_4, b) = \int_{x_3}^{x_4} f(y(x, b), y'(x, b)) dx .$$

With the help of Euler's equation (12) we see that along an extremal

$$\frac{\partial f}{\partial b} = f_y y_b + f_{y'} y_b' = y_b \frac{\partial}{\partial x} f_{y'} + y_b' f_{y'} = \frac{\partial}{\partial x} (f_{y'} y_b) ,$$

and the three partial derivatives of the function $I(x_3, x_4, b)$ have therefore the values

$$\frac{\partial I}{\partial x_3} = -f|^3 , \qquad \frac{\partial I}{\partial x_4} = f|^4 ,$$

$$\frac{\partial I}{\partial b} = \int_{x_3}^{x_4} \{f_y y_b + f_{y'} y_b'\} dx = f_{y'} y_b |_3^4 ,$$

in which the arguments of f and its derivatives are understood to be the values y, y' belonging to the family (11).

Suppose now that the variables x_3, x_4, b are functions $x_3(t), x_4(t)$, $b(t)$ of a variable t so that the end-points 3 and 4 of the extremals of the family (11) describe simultaneously

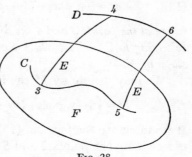

Fig. 28

two curves C and D in Figure 28 whose equations are

(13) $\qquad x = x_3(t) , \qquad y = y\big(x_3(t), b(t)\big) = y_3(t) ,$

$\qquad\qquad x = x_4(t) , \qquad y = y\big(x_4(t), b(t)\big) = y_4(t) .$

The differentials dx_3, dy_3 and dx_4, dy_4 along these curves are found by attaching suitable subscripts 3 or 4 to x, dx, and dy in the equations

(14) $\qquad dx = x'(t) dt , \qquad dy = y' dx + y_b \, db .$

From the formulas for the derivatives of I we now find the differential

$$dI = \frac{\partial I}{\partial x_3} dx_3 + \frac{\partial I}{\partial x_4} dx_4 + \frac{\partial I}{\partial b} db = f \, dx + f_{y'} \, y_b \, db \, |_3^4$$

where the vertical bar indicates the difference between the values at the points 4 and 3 of the whole expression on the right-hand side of the equation. With the help of the second of the equations (14) this gives the following important result:

The value of the integral I taken along a one-parameter family of extremal arcs E_{34} whose end-points describe the two curves C and D shown in Figure 28 has the differential

$$(15) \qquad dI = f(y, p)dx + (dy - p\,dx)f_{y'}(y, p)\,|_3^4 ,$$

where at the points 3 and 4 the differentials dx, dy are those belonging to C and D, while y and p are the ordinate and slope of E.

We may denote by I^* the integral

$$I^* = \int \{f(y, p)dx + (dy - p\,dx)f_{y'}(y, p)\} .$$

If we integrate the formula (15) between the two values of t defining the points 3 and 5 in Figure 28 we find the following useful relation between values of this integral and the original integral I.

COROLLARY 1. *For two arcs E_{34} and E_{56} of the family of extremals shown in Figure 28 the difference of the values of the integral I is given by the formula*

$$(16) \qquad I(E_{56}) - I(E_{34}) = I^*(D_{46}) - I^*(C_{35}) .$$

A region F of the plane is called a *field* if it has associated with it a one-parameter family of extremals each intersecting a curve D in one point and such that through each point (x, y) of F there passes one and but one extremal of the family. Figure 28 is a picture suggesting such a field. The function $p(x, y)$ defining the slope of the extremal of the field at a point (x, y) is called the *slope-function* of the field. With this slope-function substituted the integrand of the integral I^* depends only upon x, y, dx, dy, and the integral itself will have a well-defined value on every arc C_{35} in F having equations

$$x = x(t) , \qquad y = y(t) \qquad (t_3 \leqq t \leqq t_5)$$

of the type described on page 27. Furthermore the end-points of C_{35} determine two extremal arcs E_{34} and E_{56} of the field, and a corresponding arc D_{46}, which are related to it like those in equation (16) above. It is evident then that the value $I^*(C_{35})$ depends only upon the points 3 and 5, and not at all upon the form of the arc C_{35} joining them, since the other three terms in equation (16) have this property.

The importance of the integral I^* in the calculus of variations was first emphasized by Hilbert and it is usually called Hilbert's invariant integral.[20] Its two most useful properties are described in the following corollary:

COROLLARY 2. *For a field F simply covered by a one-parameter family of extremals all of which intersect a fixed curve D the Hilbert integral I^* formed with the slope-function $p(x,y)$ of the field has the same value on all arcs C_{35} in F with the same end-points 3 and 5. Furthermore on an extremal arc of the field I^* has the same value as I.*

The last statement follows readily since along an extremal of the field we have $dy = p \, dx$ and the integrand of I^* reduces simply to $f(y, p)dx$.

The formulas (15) and (16) are the two important ones which were mentioned in the first paragraph of this section. They remain valid in simpler forms if one of the curves C_{35} or D_{46} degenerates into a point, since then the differentials dx, dy along that curve are zero. It is useful to notice also that if one only of the arcs of the family (11) is an extremal then the formula (15) will still hold at least along that particular arc.

The results of this section, like a part of those of Section 19, page 47, are valid for the case when the integrand of the integral I is an arbitrary function $f(x, y, y')$ of the three variables x, y, y', as well as for integrands like those already considered in preceding pages in which the variable x is absent. The proofs for the more general case are word for word identical with those given above, only we must think always of the variable x as possibly being present in the function f.

39. *The envelope theorem and Jacobi's condition*. With the help of the results of the preceding section we may establish for the one-parameter family of catenaries through the point 1 another remarkable analogue of the string property of the evolute of a curve described on page 32. It is a special case of the more general theorem mentioned on page 73 as having been established for various cases by Darboux, Zermelo, and Kneser in 1894 and 1898, and which will be proved here in Section 55 of Chapter V. Lindelöf as early as 1860 had discovered a property of catenaries somewhat similar to the one in which we are interested.[21]

THE ENVELOPE THEOREM.

If two catenaries E_{14}, E_{13} of the family through the point 1

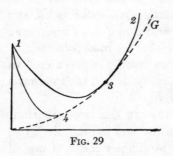

Fig. 29

touch the envelope G of the family at the points 4 and 3, as shown in Figure 29, then the areas of the surfaces of revo-

lution generated by the arcs $E_{14}+G_{43}$ and E_{13} are equal. This result may also be expressed by the equation

$$I(E_{14})+I(G_{43})=I(E_{13}) .$$

The proof is very simple with the help of the formula (16) of page 100. The family (11) of extremals is now the family of catenaries through the point 1, the curve C of the formula (16) is the fixed point 1, and the curve D is the envelope G. Hence formula (16) gives

$$I(E_{13})-I(E_{14})=I^*(G_{43}) .$$

But at every point of the arc G we have $dy=p\,dx$, where p is the slope of the catenary through that point, and hence

$$I^*(G_{43})=\int_{x_4}^{x_3}\{f(y,\,p)dx+(dy-pdx)f_{y'}(y,\,p)\}$$

$$=\int_{x_4}^{x_3}f(y,\,p)dx=I(G_{43}).$$

If this value is substituted in the preceding equation the formula of the theorem is proved.

The envelope theorem enables us to prove with ease for the catenary problem the necessary condition for a minimum which Jacobi deduced in 1837 for the general theory by a very different method. For our special case his theorem is as follows:

JACOBI'S NECESSARY CONDITION. *If a catenary arc E_{12} is to generate a surface of revolution of minimum area then the point of contact 3, shown in Figure 29, of the catenary with the envelope G of the one-parameter family of catenaries through the point 1 must not lie on the arc E_{12}.*

We see this because the arc $E_{14}+G_{43}+E_{32}$ in Figure 29 generates the same surface area as E_{12}, and we can always replace G_{43} by an arc C_{43} generating a smaller area since G_{43} is not an arc of a catenary (5). To make quite sure that G_{43} can never be such a catenary arc we note that at each point of it the equation (4) on page 91 defines a value b which is the same as the one belonging to the catenary tangent to G at that point. But these values vary from point to point of G, as shown by the second equation (6), whereas on the catenaries (5) they must be constant.

40. *The construction of a field.* If we had been studying the problem of determining surfaces of revolution of minimum area before the year 1837 when Jacobi published his necessary condition we should doubtless have concluded that every catenary arc joining the points 1 and 2 generates a minimum area. The last theorem of the preceding section shows that this would have been unjustifiable for the arc E_{132} shown in Figure 29. Similarly at the stage which we have now reached in our reasoning we might infer that the surface of revolution generated by a catenary arc joining 1 with 2 and having on it no conjugate point is smaller than that furnished by every other arc $y = y(x)$ joining 1 with 2. This conclusion would also under some circumstances be incorrect, for while a catenary arc with no conjugate point on it minimizes I with respect to other curves joining 1 with 2 and lying sufficiently near to it, there may nevertheless in some cases be curves not so near which give I a smaller value. The only way to logical safety lies in a sufficiency proof of some sort which will characterize more specifically for us the minimizing properties of our arc.

In order to make a sufficiency proof after the manner of Weierstrass it is necessary first of all to construct a field of extremals analogous to that described in Section 23, page 57, for the brachistochrone problem. Suppose that E_{12} is a catenary arc having on it no point conjugate to 1 and with the equation

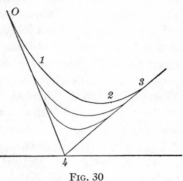

$$(17) \qquad y = b_0\, ch\frac{x - a_0}{b_0} \; .$$

Let us take a point 0 on the catenary, as in Figure 30, so near to 1 that the point 3 conjugate to 0 by the Lindelöf construction of page 94, is still at

FIG. 30

the right of 2. The tangents to the catenary at 0 and 3 meet on the x-axis at a point 4. The transformation

$$(18) \qquad x - x_4 = \frac{b_0}{b}\,(X - x_4), \qquad y = \frac{b_0}{b}\,Y$$

stretches the plane along the radii through the point 4 in such a way that every point (x, y) is replaced by a point (X, Y) whose distance from 4 is b/b_0 times that of (x, y). By substituting the values of x and y from equations (18) in the equation (17) of E_{12} we see that the points (x, y) on that catenary are transformed into points (X, Y) which satisfy the equation

$$(19) \qquad y = b\, ch\, \frac{1}{b}\left[x - x_4 + \frac{b}{b_0}(x_4 - a_0)\right] = y(x, b) \; .$$

This is also a catenary of the family $y = b \, ch[(x-a)/b]$ with the parameters $a = x_4 - b(x_4 - a_0)/b_0$ and b. If we think of b as variable we obtain therefore a one-parameter family of catenary arcs containing the original catenary E_{03} for the special value $b = b_0$. Each of these arcs is tangent to the two lines joining the point 4 to the points 0 and 3 since each is obtained from E_{03} by stretching it along the radii through the point 4.

We see readily that through each point (x, y) of the V-shaped region F bounded by the two radii joining 4 to the points 0 and 3 there passes a unique extremal of our family, and it is on account of this property that we designate F a field of extremals.

41. *Properties of the field functions.* Analytically the uniqueness of the extremal through a point of the field means that for each point (x, y) in F the equation $y = y(x, b)$ of the extremals of the field has a unique solution $b(x, y)$. We can prove without serious difficulty that this function $b(x, y)$ is continuous within and on the boundary of the V-shaped field F defined in the last section, except at the point 4, and that it has continuous derivatives in the interior of the field. The same properties are then possessed by the slope-function $p(x, y)$ of the field which is expressible analytically in terms of $b(x, y)$ in the form

$$p(x, y) = y'(x, b(x, y))$$

where the prime indicates the partial derivative of $y(x, b)$ with respect to x.

In order to prove these statements we shall need several of the partial derivatives of the function $y(x, b)$ in equation (19). If we denote derivatives with respect to x by primes, and with respect to b by subscripts, we find readily the formulas

$$y = b \, ch \, \frac{1}{b} \left[x - x_4 + \frac{b}{b_0}(x_4 - a_0) \right] = y(x, b) \ ,$$

$$y' = sh \, \frac{1}{b} \left[x - x_4 + \frac{b}{b_0}(x_4 - a_0) \right] \ ,$$

$$y_b = \frac{1}{b} [y - (x - x_4)y'] \ .$$

From the accompanying figure it is seen that y_b may be expressed in the form

$$y_b = \frac{1}{b} Y \ ,$$

where Y is the vertical distance from the point 4 to the tangent of the catenary of the field at the point (x, y). It is clear from this expression that y_b vanishes only when (x, y) is on the boundary of the field.

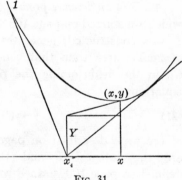

Fig. 31

Suppose now that (x, y) and $(x + \Delta x, y + \Delta y)$ are interior points of the field F and let b and $b + \Delta b$ be the corresponding values of b satisfying the equations

$$y = y(x, b), \qquad y + \Delta y = y(x + \Delta x, b + \Delta b) \ .$$

By subtracting the former from the latter and using Taylor's formula we find

(20) $$\Delta y = y' \Delta x + y_b \Delta b$$

where the arguments of the derivatives y', y_b are $x+\theta\Delta x$, $b+\theta\Delta b$ with $0<\theta<1$. By the methods used in Section 24, page 60, it follows then that $b(x, y)$ is continuous and has continuous first partial derivatives at the point (x, y).

When (x, y) is on the boundary of F the equation (20) does not furnish us with the same conclusions, since the derivative y_b vanishes on the boundary and the equation cannot be safely solved for the increment Δb. We can prove by the method used on pages 61–62, however, that $b(x, y)$ is continuous on the boundary of F elsewhere than at the vertex 4, and that at 4 it has the limit zero.

42. *The sufficiency proof.* The extremal arc E_{12} with which we started on page 105 and around which the field F was constructed, generates a surface of revolution of smaller area than that generated by every other arc C_{12} in the field joining the points 1 and 2 and having equations

$$(21) \qquad x=x(t), \qquad y=y(t) \qquad (t_1\leqq t\leqq t_2)$$

of the type described on page 27. For along such an arc the integral I and the Hilbert integral I^*, defined on page 100, formed with the integrand function $f=y(1+y'^2)^{\frac{1}{2}}$ of the catenary problem, have the values

$$I(C_{12})=\int_{t_1}^{t_2} y\sqrt{x'^2+y'^2}\,dt=\int_{s_1}^{s_2} y\,ds\,,$$

$$I^*(C_{12})=\int_{t_1}^{t_2} y\frac{x'+py'}{\sqrt{1+p^2}}\,dt=\int_{s_1}^{s_2} y\cos\theta\,ds\,,$$

where θ is again the angle at a point of the curve C_{12} between C_{12} and the extremal of the field through that point. From the equality of I^* and I on a catenary arc of the field, and the invariant property of I^*, it follows that

$$I(E_{12}) = I^*(E_{12}) = I^*(C_{12})$$

and hence that

$$I(C_{12}) - I(E_{12}) = I(C_{12}) - I^*(C_{12}) = \int_{s_1}^{s_2} y(1 - \cos\theta)ds \geq 0 ,$$

and it requires only a repetition of the argument used on page 67 to show that this difference is always positive unless C_{12} coincides with E_{12}.

We can summarize the results so far attained for our catenaries in the following theorem:

An admissible arc $y = y(x)$ $(x_1 \leq x \leq x_2)$ in the half-plane $y > 0$, joining two given points 1 and 2 and generating a surface of revolution of minimum area when rotated about the x-axis, must have the properties:

1. It is a single arc without corners of one of the catenaries $y = b\ ch[(x-a)/b]$.

2. It has on it no point of contact with the envelope G of the one-parameter family of these catenaries through the point 1.

If E_{12} is an arc having these two properties, and if F is one of the V-shaped regions shown in Figure 30, page 105, containing E_{12} in its interior and bounded by two tangents to the catenary E which meet on the x-axis, then the area of the surface of revolution generated by E_{12} is smaller than the area generated by every other arc C_{12} of the type (21) in the region F and joining the points 1 and 2.

It should be noted, as for the examples of the preceding chapters, that the theorem is more inclusive than was originally required by our problem as stated on page 90, since it holds for arcs C_{12} in parametric form as well as for admissible arcs $y = y(x)$.

43. *Solutions consisting of straight-line segments.* The theorems of the preceding sections show that no minimizing arc representable in the form $y = y(x)$ exists joining the points 1 and 2 when the point 2 is on or below the envelope G of the one-parameter family of the catenaries $y = b\ ch[(x-a)/b]$ which pass through the point 1. In the former case the one catenary joining 1 with 2 does not furnish a minimum area since it does not satisfy Jacobi's necessary condition, and in the latter case there is no such catenary. It is also of course impossible to have a minimizing arc of the form $y = y(x)$ when the points 1 and 2 are in the same vertical line, since an arc with such an equation can have but one point on each ordinate. There still remains the question therefore as to the character of the minimizing surfaces when the points 1

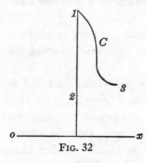

Fig. 32

and 2 lie in one of the positions which have just been mentioned.

To find an answer to this question let us consider first a segment E_{12} of the vertical line through the point 1, and an arc C_{13} with length l equal to that of E_{12}, as shown in Figure 32. Let the points at the distance s from 1 on E_{12} and C_{13}, respectively, have the ordinates y and Y. Then the difference of the areas of the surfaces of revolution generated by the two arcs is 2π times the difference

$$I(C_{13}) - I(E_{12}) = \int_0^l Y\, ds - \int_0^l y\, ds = \int_0^l (Y - y)\, ds \geqq 0.$$

The equality sign holds only if C_{13} coincides with E_{12}, since if C_{13} has a single point Y distinct from the corresponding point y of E_{12} we must have $Y > y$ and the integral on the right is then surely > 0. We have, therefore, the following theorem:

If a vertical straight line E_{12} has its upper end-point 1 in common with an arc C_{13} of the same length, as shown in Figure 32, the areas of the surface of revolution generated by rotating E_{12} about the x-axis is always less than that generated by C_{13} unless C_{13} is coincident with E_{12}.

This theorem enables us to conclude at once that if the two points 1 and 2 are in the same vertical line the straight-line segment joining them always generates a smaller surface of revolution than that generated by every other arc C_{12} with the same end-points, since such an arc C_{12} must be longer than E_{12}. But we can also deduce from it another interesting result concerning the case when the points 1 and 2 are not in the same vertical.

Let 3 and 4 be the points on the x-axis below 1 and 2, respectively, as shown in Figure 33. If an arc C_{12} in the half-plane $y \geqq 0$ has length greater than $y_1 + y_2$ then the theorem above shows that area of the surface of revolution generated by C_{12} is always greater than

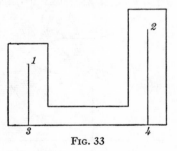

FIG. 33

or equal to that generated by the broken line L_{1342} consisting of the ordinates of the points 1 and 2 and the segment of the x-axis which they intercept. If we take a neighborhood of this broken line, of the type shown in Figure 33, so small that every arc C_{12} in it joining 1 with 2 has necessarily a length greater than $y_1 + y_2$, then it is clear that at least in that neighborhood the line L_{1342} is a minimizing arc for our problem of determining a curve joining the points 1 and 2 and generating a surface of revolution of minimum area. It is the Goldschmidt discontinuous solution referred to on page 88 of the introductory section of this chapter.

44. *A second type of field.* It is evident that the V-shaped field F in which the catenary arc E_{12} has been proved to furnish a minimum is not unique, for the two points 0 and 3 can be moved slightly to the right or left without destroying the properties needed for the field. It is not surprising, therefore, to find that there is a field of a still different type in which the catenary E_{12} retains its minimizing properties. We can, in fact, prove the following theorem:

If E_{12} is a catenary of the family $y = b\, ch[(x-a)/b]$ having on it no point conjugate to 1 except possibly at 2,

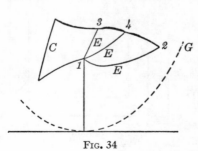

FIG. 34

then the surface of revolution which it generates is smaller than that generated by every other arc C_{12} with equations of the type (21) on page 108, joining 1 with 2, and, except possibly at 2, lying entirely above the envelope G of the one-parameter family of catenaries through the point 1 shown in Figure 34.

In order to prove this let us note in the first place that through each point 3 above the envelope G and distinct from 1 there passes a unique extremal E_{13} containing no point of contact with G, as shown in Figure 34. The vertical straight line through 1 is now understood to be included among the extremals in order to make this statement true for every point of the field. In the next section we shall see that the value $I(E_{13})$ of our integral varies continuously with the point 3 and approaches the value zero when 3 approaches the point 1.

Consider now an arbitrarily selected arc C_{12} above the envelope G. From the results of the last two sections we know that when a point 4 of this arc is first chosen and a second point 3 of the arc afterward selected sufficiently near to it, the value $I(E_{13}+C_{34})$ exceeds $I(E_{14})$ unless the arc $E_{13}+C_{34}$ coincides with E_{14}. Since

$$[I(C_{14})-I(E_{14})]-[I(C_{13})-I(E_{13})]=I(E_{13}+C_{34})-I(E_{14})$$

it follows that as the point 3 traverses the arc C_{12} from the point 1 to the point 2 the difference $I(C_{13})-I(E_{13})$ starts at the value zero and does not decrease, so that we have $I(C_{12}) \geqq I(E_{12})$.

The equality sign can hold only when C_{12} coincides with E_{12}. For if there is a point 3 on C_{12} and not on E_{12} then there is also a point 4 between 3 and 2 on C_{12} such that E_{14} is distinct from both E_{13} and E_{12}, and furthermore such that 4 is the first point at which C_{34} meets E_{44}. If we now take 3 sufficiently near to 4 we shall have $I(E_{13}+C_{34})>I(E_{14})$ and hence the difference $I(C_{13})-I(E_{13})$ actually increases as 3 traverses C_{12}, so that $I(C_{12})>I(E_{12})$.

45. *The continuity of the extremal integral.* It is easy to see that the value $I(E_{13})$ of our integral approaches zero as 3 approaches the point 1. For there is in no case a point conjugate to 1 on E_{13} and the straight-

line segment L_{13} joining 1 and 3 always lies in a V-shaped field for E_{13} like that described in Section 40, page 104. Hence we have $I(E_{13}) < I(L_{13})$ and $I(L_{13})$ evidently approaches zero as the point 3 approaches 1.

If a point 3 of the field is not on the ordinate $x = x_1$ the value of the extremal integral along the catenary arc E_{13} of the field is

$$I(E_{13}) = \int_{x_1}^{x_3} y(x, a)\sqrt{1 + y'^2(x, a)}\, dx\, .$$

Here $y(x, a)$ is the function (7) of page 92 defining the one-parameter family of catenaries through the point 1, and a is the single-valued function of x_3 and y_3 satisfying the equation $y_3 = y(x_3, a)$. We may prove by the methods of Section 24, page 60, that the function $a(x_3, y_3)$ satisfying this equation is continuous at all points 3 in the interior and on the boundary of the field F which we are considering, except those on the line $x = x_1$. Hence the integral $I(E_{13})$ also varies continuously at points 3 which are not on this line.

Consider finally the case when the point 3 lies on the ordinate $x = x_1$. We wish to show that $I(E_{14})$ approaches $I(L_{13})$ when the point 4 approaches 3. From the results of preceding sections we know that the surface generated by the vertical line L_{13} is smaller than that generated by every other arc joining its end-points, and hence, with the help of the remarks in the first paragraph of this section, that

$$I(L_{13}) \leqq I(E_{14} + L_{43}) \leqq I(L_{14} + L_{43})\, .$$

But from this result we deduce readily the inequalities

$$-I(L_{43}) \leqq I(E_{14}) - I(L_{13}) \leqq I(L_{14}) - I(L_{13})\, .$$

Since the first and last members approach zero as the point 4 approaches the point 3, it follows that the second expression must do the same.

The continuity of the integral $I(E_{13})$ as the point 3 varies on the interior and boundary of the field F, which was assumed without proof in the reasoning of the last section, is thus established for all cases.

46. *The absolute minimum.* Since we know that the catenary arcs E_{12} without contacts with G, and the Gold-schmidt discontinuous so-lution L_{1342}, in Figure 35, both furnish minima with respect to curves lying near them, it is reasonable to ask whether or not one of them furnishes a minimum when compared with all the arcs C_{12} joining 1 with 2 in the half-plane $y \geqq 0$. An arc which has this property is said to furnish an abso-lute minimum for our

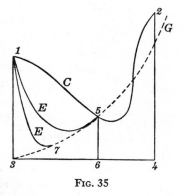

Fig. 35

problem. To answer this question we can first prove the following statement:

Every arc C_{12} distinct from L_{1342} and having a point 5 in common with the envelope G, as shown in Figure 35, gener-ates a larger surface of revolution than the Goldschmidt solution.

Let 5 be the first point at which C_{12} has an intersec-tion with G. Then we may conclude that $I(C_{15}) \geqq I(L_{1365})$. For when a point 7 on G is sufficiently near to 3 the length of $E_{17} + G_{75}$ is greater than $y_1 + y_5$, and hence from the theorems of pages 112 and 102, and the last paragraph of Section 43 on page 112, we have successively

$$I(C_{15}) \geqq I(E_{15}) = I(E_{17} + G_{75}) \geqq I(L_{1365}) .$$

The equality signs all hold only when 5 is at the point 3 and C_{15} coincident with L_{13}. Furthermore if we let the point 5 move along C_{12} toward the point 2 the difference

$$I(C_{15}) - I(L_{1365}) = \int_0^{s_5} y \ ds - \tfrac{1}{2}(y_1^2 + y_5^2) ,$$

where s is the length of arc measured along C_{15}, has its derivative $y_5(1 - dy_5/ds_5)$ with respect to s_5 always positive or zero since the absolute value of the ratio dy_5/ds_5 never exceeds unity. Hence the difference is never decreasing as 5 moves toward 2 on the arc C_{12}, and since it starts with a positive or zero value when 5 is on G, we infer that when 5 has arrived at the point 2 we have $I(C_{12}) - I(L_{1342}) \geqq 0$. One can verify readily that the equality sign holds only if C_{12} and L_{1342} are identical. It is now also clear that

When there are fewer than two catenaries joining the points 1 and 2 the Goldschmidt solution always furnishes an absolute minimum.

For in that case the point 2 is on or below the envelope G and every curve joining 1 with 2 must intersect G.

When 2 is above G the catenary E_{12} having no contact with G, and the Goldschmidt solution L_{1342}, both furnish minima in sufficiently small neighborhoods, and the one which generates a smaller area than the other surely provides an absolute minimum.

For example, in the case when $I(E_{12}) < I(L_{1342})$, then $I(E_{12})$ is smaller than all the values of I on arcs C_{12}

above the envelope G, by the theorem of page 112, and smaller than the values $I(C_{12})$ for curves meeting G since for such curves $I(E_{12}) < I(L_{1342}) \leqq I(C_{12})$. The argument is quite similar for the other case when $I(L_{1342}) < I(E_{12})$. When the two are equal then each of the arcs E_{12} and L_{1342} generates a smaller surface of revolution than other arcs with the same end-points.

There is an interesting geometric criterion, first presented by MacNeish,[22] for determining which of the values $I(E_{12})$ and $I(L_{1342})$ is the smaller. The difference between these values is

$$I(E_{12}) - I(L_{1342}) = \int_0^{s_2} y \, ds - \tfrac{1}{2}(y_1^2 + y_2^2) \ .$$

As the point 2 moves from 1 along a fixed catenary E the derivative $y_2(1 - dy_2/ds_2)$ of this difference is positive since the tangent to the catenary is never vertical and the absolute value of the derivative dy_2/ds_2 is therefore never as great as unity on it. When 2 is at 1 the difference $I(E_{12}) - I(L_{1342})$ is negative since then $I(E_{12}) = 0$; and when 2 is on G it is positive since $I(L_{1342})$ is smaller than the value of I on every other curve intersecting G. It follows then that $I(E_{12}) = I(L_{1342})$ for one position only of the point 2 on the catenary E between these extremes. MacNeish determined the character of the locus of such points 2 by methods analogous to those which had been used in the determination of the form of the envelope G. In terms of the parameter $u = a + (x - x_1) \, ch\, a/y_1$ the equations of the family (7) of catenaries through the point 1, on page 92, may be written

$$(22) \qquad x = x_1 + \frac{y_1}{ch\ a}(u - a), \quad y = y_1 \frac{ch\ u}{ch\ a},$$

and the values of $I(E_{12})$ and $I(L_{1342})$ in terms of the parameter u of the point 2 are readily found to be

$$I(E_{12}) = \int_a^u y \sqrt{x_u^2 + y_u^2}\, du = \tfrac{1}{2}\left(\frac{y_1}{ch\ a}\right)^2 [u + sh\ u\ ch\ u]_a^u ,$$

$$I(L_{1342}) = \tfrac{1}{2}(y_1^2 + y_2^2) = \tfrac{1}{2}\left(\frac{y_1}{ch\ a}\right)^2 [ch^2\ a + ch^2\ u] .$$

These two are equal when

$$u + sh\ u\ ch\ u - ch^2\ u = a + sh\ a\ ch\ a + ch^2\ a .$$

This equation and the equations (22) of the catenary arc define the locus H of the points where $I(E_{12}) = I(L_{1342})$. MacNeish discussed the form of the curve H and plotted

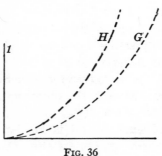

it from numerical data. It turns out that its shape is similar to that of the envelope G, as shown in Figure 36. With the help of these curves our results for absolute minima may be described as follows:

For a point 2 above the curve H in Figure 36 the Goldschmidt discontinuous so-

Fig. 36

lution L_{1342} joining 1 to 2 generates a minimum surface of revolution relative to those generated by other arcs of the type (21) on page 108 joining the same two points and lying in a sufficiently small neighborhood of L_{1342}; but the smallest surface of all, the absolute minimum, is in this case furnished by the unique catenary arc E_{12} joining 1 with 2 and having on it no contact point with the envelope G. When 2 is on H the surfaces generated by L_{1342} and E_{12} are equal in area and smaller than those generated by other arcs joining these two points.

When 2 is between H and G the catenary arc furnishes a relative minimum and the Goldschmidt solution the absolute minimum. When 2 is on or below G the Goldschmidt solution is the only minimizing arc joining 1 with 2 and it furnishes an absolute minimum.

47. *Soap films.* It has already been remarked on page 7, that the problem of determining the form of a soap film stretched between two circles whose planes are parallel and whose centers are on a common axis perpendicular to their planes, is equivalent to that of determining a curve joining two given points and generating a surface of revolution of minimum area. We have seen that at least when the circles are sufficiently near to each other the meridian curve of the surface must be a catenary. If the right-hand circle in Figure 2, page 7, is moved slowly away from the other in the direction of their common axis the meridian catenary takes successive positions in the one-parameter family of catenary arcs through the point 1. When a certain point has been reached the surface becomes unstable. Without any further movement of the circles it will gradually contract and separate into two portions which retire into the planes of the circles to form two circular disks, the Goldschmidt discontinuous solution of our minimum problem. The moment at which this separation takes place is the one when the point 2 reaches in its horizontal movement the envelope G of the family of catenaries through the point 1.

One might expect that the instability would occur at the instant when 2 reaches the locus H on which the catenary solution and the Goldschmidt solution give equal areas. At that point, however, the catenoid surface is still a minimum with respect to other surfaces in its immediate neighborhood stretched between the two circles, and the only effect of a slight disturbance is to cause it to oscillate back into its original position.

The equations which determine the relative position of the two circles at the moment of instability are transcendental in character, but it is not difficult to secure numerical data of sufficient accuracy for comparison with those obtained by experiment. If we use the parameter u again the envelope G is the locus of the points (x, y) determined by the three equations

$$(23) \quad \frac{x - x_1}{y_1} = \frac{u - a}{ch\, a}, \qquad \frac{y}{y_1} = \frac{ch\, u}{ch\, a}, \qquad u - \frac{ch\, u}{sh\, u} = a - \frac{ch\, a}{sh\, a}.$$

The first of these is in another form the equation defining u; the second is obtained from equation (7), page 92;

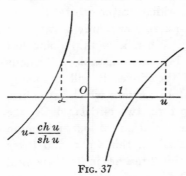

FIG. 37

and the last is equivalent to the condition $y_a = 0$ along G and is obtained by equating to zero the expression for y_a given on page 93 after inserting the values $y' = sh\, u$, $y_1' = sh\, a$ and the values of x and y from the first two of the equations (23). If we plot carefully the function $u - chu/shu$, then for every negative value a a positive value u can be measured satisfying the third equation (23), as shown in Figure 37. The corresponding point (ξ, η) defined by the equations

$$\xi = (u - a)/ch\, a, \quad \eta = ch\, u/ch\, a$$

describes the particular envelope G which is plotted to scale in Figure 25, page 86, for the point $(x_1, y_1) = (0, 1)$. In order to find approximately, in a more general case, the difference $x - x_1$ of the abscissas of 1 and its conjugate (x, y) we have only to find the value ξ corresponding to the value $\eta = y/y_1$ on the graph of G in Figure 25 and then set $x - x_1 = \xi y_1$. Still greater accuracy could be secured if desired by using methods of approximation in the solution of equations (23) for u, a, $x - x_1$ when y_1 and y are given.

Professor Mary E. Sinclair has studied in a very interesting way a modification of this soap-film problem.[23] If an inverted funnel is set in a larger funnel and moistened with soap solution, then when the smaller one is withdrawn in the direction of its axis a surface of revolution is formed whose cross-section with those of the funnels is shown in Figure 38. It is a catenoid one of whose

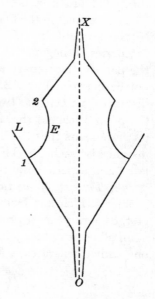

Fig. 38

bounding circles is always the greatest circle of the smaller funnel, while the other slides up or down the inner surface of the larger funnel as the distance between the two funnels is altered. At their intersection the soap film and the larger funnel meet at right angles. As the surface is elongated by enlarging the distance between the two funnels a certain point is reached when the sur-face becomes unstable and separates as before.

The analytical problem corresponding to this experi-ment is that of finding among the curves joining a given point 2 to a given straight line L a particular one which when rotated about the axis OX will generate a minimum area. The curve must in the first place be a catenary orthogonal to the straight line L. The one-parameter family of catenaries orthogonal to L has an envelope G touching the particular catenary arc E_{12} in a point 3 which must not lie between 1 and 2. Miss Sinclair determined the distance between the two funnels at the moment when the point 2 reaches the envelope G and verified by actual measurements that the film becomes unstable when that distance is reached. The agreement between her calcu-lated and experimental results was surprisingly close.

48. *The case of one variable end-point.* The soap-film problem which has just been described is a special case of the more general problem of determining among the admissible arcs which join a fixed point 1 to a fixed curve N one which generates a minimum surface of revolution.

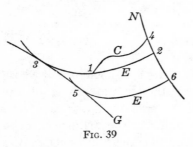

Fig. 39

The analysis which en-ables us to solve this problem is similar to that of Sections 29–31, pages 70–77, concern-ing the determination of a path of quickest descent from a point to a curve, and it will be sufficiently intelligible if presented more concisely.

The minimizing arc E_{12} in Figure 39 must in the first

place furnish a minimum surface of revolution when compared with other arcs joining the points 1 and 2, since every such arc also joins 1 with N. Hence it must be a catenary arc without corners.

It must furthermore be cut at right angles by N. For there is a one-parameter family of arcs $y = y(x, b)$, joining 1 with N and containing E_{12} for a special parameter value $b = b_0$, as will be shown explicitly in the next paragraph. The values of I taken along the members of this family have a differential along E_{12} which is given by the formula (15) of page 100 when the curve C of that formula is replaced by the point 1 and D by N, attention being paid to the remark on page 102 which explains that the formula is still applicable when E_{12} is the only extremal in the family. For the special integrand function $f = y(1 + y'^2)^{\frac{1}{2}}$ of this chapter the formula (15) just referred to gives the value

$$dI = f\,dx + (dy - y'dx)f_{y'} \mid^2 = y\,\frac{dx + y'dy}{\sqrt{1 + y'^2}} \mid^2$$

for the differential of I along E_{12}. Since this must vanish when $I(E_{12})$ is a minimum it follows that the direction $dx : dy$ of N is necessarily perpendicular to the direction $1 : y'$ of E_{12} at the point 2.

To show that there is a one-parameter family of curves $y = y(x, b)$ of the kind used in the last paragraph suppose that N has parametric equations of the forms

$$x = g(b) , \qquad y = h(b)$$

and that it intersects E_{12} at the point 2 for the parameter value b_2. If the equation of E_{12} is $y = y(x)$ then the family

$$y = y(x) + \frac{h(b) - y(g(b))}{g(b) - x_1}(x - x_1) = y(x, b)$$

contains E_{12} for the parameter value $b = b_2$, and all of its arcs pass through the point 1 for $x = x_1$ and intersect N for $x = g(b)$.

We may construct a one-parameter family of the catenaries $y = b\ ch[(x-a)/b]$ cutting N at right angles, by the method described on page 81 for curves of the family $y = b\phi[(x-a)/b]$. The point 3 where the envelope G of this family touches the extension of E_{12} in Figure 39 is called *the focal point of N on E_{12}*. The formula (16) on page 100 justifies as on page 73 an envelope theorem which says that $I(G_{35}) + I(E_{56}) = I(E_{32})$ for every position of the point 5 on G, and by means of it we also prove that the focal point 3 can never lie on the arc E_{12} if E_{12} is to furnish a minimum.

By an argument similar to that of Section 31 on page 75, we show that when the focal point 3 does not lie on E_{12} the one-parameter family of catenaries orthogonal to N simply covers a field near E_{12}, and we prove that in this field E_{12} does actually furnish a minimum. We have then the following results for this problem:

Let a fixed point 1 and a fixed curve N be given. An admissible arc E_{12}, as shown in Figure 39, page 122, which generates a minimum surface of revolution as compared with those generated by other admissible arcs joining 1 with N, must have the following properties:

1. *It is a catenary of the family $y = b\ ch[(x-a)/b]$.*

2. *It is cut at right angles by the curve N at their intersection point 2.*

3. *It has on it no contact point with an envelope G of a one-parameter family of catenaries orthogonal to the curve N and containing it as a member.*

If an arc E_{12} joins 1 to N and has these properties then *it has a neighborhood F such that the surface of revolution generated by E_{12} is smaller than that generated by every other arc C_{14} of the type (21) on page 108 lying in F and joining N with 1.*

The theorem needs alteration, as has been true for similar theorems on pages 33 and **77,** in the exceptional case when the envelope G has no branch projecting toward **2** at its contact point **3** with the curve E_{12}.

49. *The geometrical construction for the focal point.* The family of catenaries $y = b\,ch[(x-a)/b]$ which furnishes the minimizing arcs for the problem of determining surfaces of revolution of minimum area, is a special case of the family $y = b\phi[(x-a)/b]$ of curves considered in Section 33, page 79. We must expect then that the geometric construction there described for the focal point of a curve will again be applicable. It is not necessary to repeat the analysis which led to that construction, but it may be of interest to have the figure for the extremals of

the present chapter since it has an appearance quite different from that for the cycloids.

Let N in Figure 40 be the curve whose focal point on the catenary E intersecting it at right angles at the

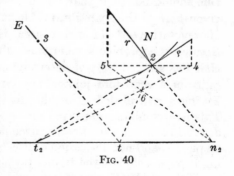

FIG. 40

point 2 is to be determined. Draw the radii of curvature ρ and r of N and E at the point 2. Join the end-points 4 and 5 of their projections on a parallel to the x-axis through 2 to the intersections t_2 and n_2, respectively, with the x-axis of the tangent and normal to the catenary E at the point 2. The lines so drawn meet in a point 6, which when joined to 2 determines the point t on the x-axis. The tangent to the catenary from t determines the focal point 3 of N on E. Evidently from this construction we have

$$\frac{n_2-t}{t-t_2}=\frac{\overline{52}}{\overline{24}}$$

which is the characteristic property of the focal point specified by equation (38) of page 83.

The construction just given is a slight modification of that of Professor Mary E. Sinclair, who was the first person to devise a geometric construction for a focal point of a curve N on a catenary analogous to the Lindelöf construction for the conjugate point.

50. *Further remarks concerning the catenary problem.* One should not infer that the discussion which has been given here of the soap-film problems leading to surfaces of revolution of minimum area is at all complete. Besides the problems mentioned above Professor Sinclair has studied that of determining the form and stability of a soap film joining two circles C_1 and C_2 and having a central disk D of film, as shown in Figure 41. This is a configuration often met experimentally in endeavoring to get the simpler catenoid surface described on page 7, and her discussion of it is of the same interesting type as that for the funnel problem of page 121.

If the circles C_1 and C_2 are solid disks with a hole in one of them through which air can be blown into the space inclosed by the film and the disks, then the problem suggested is that of determining the form of a soap film which incloses with the disks a given amount of air.

This is analytically the problem of determining among the curves joining two given points, and generating solids of revolution of given volume when rotated around the x-axis, one which at the same time generates a surface of revolution of minimum area. The curves which may give a minimum or a maximum for this problem are well known. They are the so-called elastic curves generated by the foci of ellipses and hyperbolas which roll upon the x-axis. The ones generated by the foci of ellipses are usually called unduloids, and those generated by hyperbolas nodoids. The analysis of these curves is more compli-

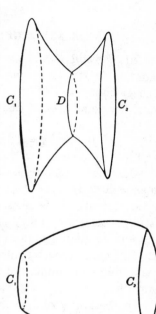

Fig. 41

cated than that of the catenary, and no elementary method of constructing the conjugate and focal points, such as has been described above for cycloids and catenaries, has ever been discovered.

CHAPTER V

A MORE GENERAL THEORY

51. *Formulation of the problem.* We have been considering in the preceding chapters several problems of the calculus of variations whose integrals to be minimized were all special cases of the more general integral

$$(1) \qquad I = \int_{x_1}^{x_2} f(x, y, y')dx$$

in which the integrand is allowed to contain the variable x as well as the variables y and y' which have hitherto been present. It is clear that if we can find characteristic properties of minimizing arcs for this integral we shall have results applicable not only to the problems already considered but also to a much larger variety of maximum and minimum questions of the calculus of variations.

In the study of this more general problem we shall need to have a class of so-called *admissible arcs* of the form

$$(2) \qquad y = y(x) \qquad (x_1 \leqq x \leqq x_2)$$

on each of which the integral I has a well-defined value, and our problem will then be to find among all the admissible arcs joining two given points 1 and 2 one which minimizes the integral I.

The definition of the class of admissible arcs may be made in many ways, each of which gives rise to a distinct problem of the calculus of variations. For a special problem the properties defining the class will in general be in part necessitated by the geometrical or mechanical character of the problem itself, and in part free to be chosen with a large degree of arbitrariness. An example of a property of the former type is the restriction for the brachistochrone problem that the curves considered shall all lie below the line $y = a$, since on arcs above that line the integral expressing the time of descent has no meaning. On the other hand we frequently find it convenient to make the arbitrary restriction that our curves shall all lie in a small neighborhood of a particular one whose minimizing properties we are investigating; and we may specify with considerable freedom the continuity properties of the class of curves in which we wish to seek a minimizing arc, always remembering that on each of the arcs of our class the integral I must have a well-defined value.

In order to make a definition of a class of admissible arcs which will be generally applicable let us first assume that there is a region R of sets of values (x, y, y') in which the integrand function $f(x, y, y')$ of the integral (1) is continuous and has continuous derivatives of as many orders as may be needed in our theory. For all ordinary purposes it will be sufficient if it has continuous partial derivatives at least up to and including those of the fourth order. The sets of values (x, y, y') interior to the region R may for convenience be designated as *admissible sets*. An arc (2) will now be called an *admissible arc* if it is continuous and has a continuously turning tangent except possibly at a finite number of corners, and if the sets of values $(x, y(x), y'(x))$ on it are all admissible according to the definition just given. For an admissible arc the interval x_1x_2 can always be subdivided into a number of partial intervals on each of which $y(x)$ is continuous and has a continuous derivative. At a value x where the curve has a corner the derivative $y'(x)$ has two values which we may denote by $y'(x-0)$ and $y'(x+0)$, corresponding to the backward and forward slopes of the curve, respectively.

52. *A summary of results.* In the statement of the following conditions we shall need to use the function

$$E(x, y, y', Y') = f(x, y, Y') - f(x, y, y') - (Y'-y')f_{y'}(x, y, y')$$

which was introduced by Weierstrass and which is called the Weierstrass E-function. Its form is easy to remember if we notice that it is $f(x, y, Y')$ minus the first two terms of the expansion of this function by Taylor's formula in powers of $(Y'-y')$.

There are in all four conditions which will be proved in later sections to be necessary for a minimum and which are stated here without proof in order that the reader may have in advance some idea of the purposes of this chapter. With the understanding that the equation of the minimizing arc E_{12} is $y=y(x)$, the first three of these are as follows:

I. *For every minimizing arc E_{12} there exists a constant c such that the equation*

$$(3) \qquad f_{y'}(x, y(x), y'(x)) = \int_{x_1}^{x} f_y(x, y(x), y'(x))dx + c$$

holds identically on E_{12}.[24] An immediate consequence of this equation is that on each arc of E_{12} having a continuously turning tangent Euler's differential equation

$$(4) \qquad \frac{d}{dx}f_{y'} - f_y = 0$$

must also be satisfied.

II (*Weierstrass*). *At every element* (x, y, y') *of a minimizing arc* E_{12} *the condition*

$$E(x, y, y', Y') \geqq 0$$

must be satisfied for every admissible set (x, y, Y') *different from* (x, y, y').

III (*Legendre*). *At every element* (x, y, y') *of a minimizing arc* E_{12} *the condition*

$$f_{y'y'}(x, y, y') \geqq 0$$

must be satisfied.

The solutions $y = y(x)$ of Euler's differential equation (4) which are admissible arcs and have furthermore continuous first and second derivatives, are called *extremals*,
and we shall see that through a fixed point 1 there passes in general a one-parameter family of such curves. If such a family has an envelope G, as

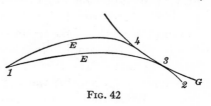

Fig. 42

shown in Figure 42, then the contact point 3 of an extremal arc E_{12} of the family with the envelope is called a *point 3 conjugate to 1 on* E_{12}.

We shall see that there is an envelope theorem for the general theory considered in this chapter which has as a special case the theorem described on page 102 for catenaries. It says that the value of the integral I taken along the composite arc $E_{14} + G_{43}$ in the figure is equal to its value along the arc E_{13} for all positions of the point 4 preceding 3 on G. With the help of this theorem we shall prove the following fourth necessary condition:

IV (*Jacobi*). *On a minimizing extremal arc E_{12} with $f_{y'y'} \neq 0$ everywhere on it there can be no point 3 conjugate to 1 between 1 and 2.*

The order of discovery of these conditions was I, III, IV, II with the exception that Euler's differential equation was not originally deduced by means of the equation (3). It happens, however, that Legendre's condition is a very easy consequence of that of Weierstrass, and for this reason the order indicated by the Roman numbers is more convenient for us than the historical one.

Weierstrass made a further very important contribution to the theory of the calculus of variations when he proved that certain sets of conditions are sufficient to insure the minimizing property for a particular arc. Up to his time students of the theory seem to have tacitly assumed, after the discovery of each new necessary condition, that the conditions then known were sufficient for a minimum. The fact that new conditions had appeared one after another was apparently conclusive evidence to Weierstrass that a sufficiency proof was needed.

Let us with him define a *field*, as in the preceding chapters, to be a region F of the plane which has associated with it a one-parameter family of extremal arcs each of which inter-

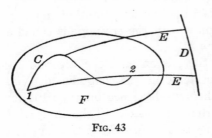

FIG. 43

sects once a curve D and which have the further property that through each point (x, y) of F there passes

one and but one extremal of the family. Such a field is
illustrated in Figure 43. Let us furthermore designate
the function $p(x, y)$ defining the slope of the extremal of
the field at a point (x, y) as the *slope-function* of the field.
The following theorem, which will be proved later, is then
fundamental for all of the sufficiency proofs:

THE FUNDAMENTAL SUFFICIENCY THEOREM. *Let E_{12}
be an extremal arc of a field F such that at each point (x, y)
of F the inequality*

$$(5) \qquad E(x, y, p(x, y), y') \geqq 0$$

*holds for every admissible set (x, y, y') different from (x, y, p).
Then $I(E_{12})$ is a minimum in F, or, more explicitly, **the
inequality $I(E_{12}) \leqq I(C_{12})$ is satisfied for every admissible
arc C_{12} in F joining the points 1 and 2. If the equality
sign is excluded in the hypothesis (5) then $I(E_{12}) < I(C_{12})$
unless C_{12} coincides with E_{12}, and the minimum is a so-called
proper one.**

For a number of special problems this theorem is
very powerful. When we sought to find the arc of
shortest length joining two points 1 and 2, for example,
we found that the straight line E_{12} joining them is an
extremal of a field which consists of the whole plane
covered by the straight lines parallel to E_{12}, and one may
readily verify that the condition (5) for the integrand
function of that problem holds without the equality
sign in such a field. Similarly a cycloid arc E_{12} for the
brachistochrone problem is an extremal of a field consisting of the half-plane below the line $y = a$ covered by the
cycloids concentric with E_{12}, and the stronger condition
(5) holds also in that case. For each of these problems,

then, an extremal arc E_{12} will furnish a minimum in a very large field.

In general we cannot hope to find such extensive fields surrounding a given extremal, as is indicated by the catenary problem, but we shall see that an arc E_{12} which has suitable properties will at least be an extremal arc of a field of limited extent. Using a notation introduced by Bolza let us designate by II', III' the conditions II, III with the equality sign excluded, and by IV' the condition IV when strengthened to exclude the possibility of a conjugate point at the end-point 2 as well as between 1 and 2 on E_{12}. It will be proved in a later section that for an extremal arc E_{12} which satisfies the conditions I, III', IV' there is always some neighborhood F which is a field simply covered by a one-parameter family of extremals having E_{12} as a member of the family.

The value $I(E_{12})$ is said to be a *weak relative minimum* if there is a neighborhood R' of the values (x, y, y') on E_{12} such that the inequality $I(E_{12}) \leq I(C_{12})$ is true, not necessarily for all admissible arcs C_{12}, but at least for all those whose elements (x, y, y') lie in R'. With the help of the sufficiency theorem stated above and the field described in the last paragraph we shall be able to prove that an arc E_{12} which satisfies the conditions I, III', IV' will surely make the value $I(E_{12})$ at least a weak relative minimum. This result will be established by replacing the original region R by R' and choosing R' so small that every admissible arc with respect to it is necessarily in the field F, and furthermore so small that the condition (5) of the theorem holds in F in its stronger form with respect to all of the sets (x, y, y') in R'.

Following Bolza again let us denote by II_b the condition II strengthened to hold not only for elements (x, y, y') on E_{12} but also for all such elements in a neighborhood of those on E_{12}. It will be proved that for an arc which satisfies the conditions I, II_b', III', IV' the field $\bar{F}$ about E_{12}, existent as a result of the conditions I, III', IV', can be so restricted in size that the stronger condition (5) holds in it with respect to the sets (x, y, y') in the region R itself. The value $I(E_{12})$ will therefore again be a minimum in F, and it is called a *strong relative minimum* because it is effective with respect to all admissible comparison curves C whose elements (x, y, y') have their points (x, y) in a small neighborhood F of those on E_{12}. No restrictions are in this case imposed upon the slopes y' except those due to the definition of the original region R.

There is a simpler set of conditions for a strong relative minimum in the case when the region R has the property that the element (x, y, y') is always in R whenever y' lies between y'_1 and y'_2 and (x, y, y'_1) and (x, y, y'_2) are both in R. Let III_b be the condition III strengthened to hold for all admissible elements (x, y, y') having their points (x, y) in a neighborhood of those on E_{12}. It will be proved that II_b' is a consequence of III_b' when R has the property described above, and it follows at once from the result stated in the last paragraph that the conditions I, III_b', IV' are also sufficient to insure a strong relative minimum.

So far the remarks made in this section have applied only to minima and not to maxima. It will be evident in later sections, however, that the conditions for a minimum become analogous ones for a maximum if the inequality signs in the conditions are changed in sense wherever they occur. We can also modify our problem by seeking to find a minimizing or maximizing arc in the class of admissible arcs joining a fixed point and a fixed curve, or two fixed curves, instead of two fixed points. The results for these two problems are described somewhat briefly in Sections 63–65 below.

53. *The first necessary condition and two fundamental formulas.* The proof which is required for the first necessary condition need not be repeated here since it is precisely that of Section 19, page 47. The presence of the variable x in the integrand of our integral does not interfere in any way with the validity of the deduction there made of the equations

$$(6) \qquad f_{y'} = \int_{x_1}^{x} f_y \, dx + c \, , \qquad \frac{d}{dx} f_{y'} - f_y = 0$$

on a minimizing arc. An extremal is by definition an admissible arc with continuous first and second derivatives satisfying these equations, and along it the Euler equation can always be written, for the general problem of this chapter, in the form

$$(7) \qquad \frac{d}{dx} f_{y'} - f_y = f_{y'x} + f_{y'y} y' + f_{y'y'} y'' - f_y = 0 \, .$$

It is a differential equation of the second order since the derivative y'' is the highest which it contains.

The arguments of Section 38, page 98, are also applicable at once to the general integral which we are now considering. If an arc E_{24} varies so that its end-

points describe simultaneously two curves C and D then
the differential of the value of the integral I along it is

$$(8) \qquad dI(E_{34}) = f(x, y, p)dx + (dy - p\, dx)f_{y'}(x, y, p) \Big|_3^4$$

at every position in which it is an extremal.   This result
holds also at every position in which it satisfies the equations (6), as a glance at the proof in Section 38 will show.
The differentials dx, dy in the last formula are those of
the curves C and D at the points 3 and 4 of Figure 28
on page 99, and the values to be inserted for p in the
difference indicated are the slopes of E_{34} at these two
points.

If the variable arc E_{34} is always an extremal, then the
difference between the values of I on the arcs E_{34} and
E_{56} of Figure 28 is

$$(9) \qquad I(E_{56}) - I(E_{34}) = I^*(D_{46}) - I^*(C_{35}),$$

where I^* is the Hilbert integral

$$(10) \quad I^* = \int \left\{ f(x, y, p)dx + (dy - p\, dx)f_{y'}(x, y, p) \right\}.$$

If the slope-function $p(x, y)$ of a field such as was described on page 101 is substituted for p in the integrand
of this integral, then along all admissible arcs in the
field having the same end-points the values of I^* are the
same, as was proved readily on that page with the help
of the formula (9) just given.   Furthermore on an extremal arc of the field the value of I^* is the same as that
of I since along such an arc $dy = p\, dx$.

The formulas (8) and (9) and the properties of the
Hilbert integral will be of frequent service for the general

theory, as they were for the more special problems of the preceding chapters.

54. *The necessary conditions of Weierstrass and Legendre.* In order to prove Weierstrass' necessary condition let us select arbitrarily a point 3 on our minimizing arc E_{12}, and a second point 4 of this arc so near to 3 that there is no corner of E_{12} between them. Through the

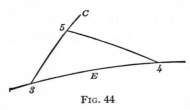

Fig. 44

point 3 we may pass an arbitrary curve C with an equation $y = Y(x)$, and the fixed point 4 can be joined to a movable point 5 on C by a one-parameter family of arcs E_{54} containing the arc E_{34} as a member when the point 5 is in the position 3. We shall see presently that such a family can very easily be constructed. If the integral $I(E_{12})$ is to be a minimum then it is clear that as the point 5 moves along C from the point 3 the integral

$$(11) \qquad I(C_{35}+E_{54}) = \int_{x_3}^{x_5} f(x, Y, Y')dx + I(E_{54})$$

must not decrease from the initial value $I(E_{34})$ which it has when 5 is at the point 3. Evidently at the point 3 the differential of this integral with respect to x_5 must not be negative.

The differential of the term $I(E_{54})$ in the expression (11), at the position E_{34}, is given by the formula (8) of the preceding section when the curve D of that formula is replaced by the fixed point 4 at which $dx_4 = dy_4 = 0$. For the formula (8) holds along every arc of the family in question which satisfies the equations (6), and we know already that our minimizing arc must satisfy these equations. Since the derivative of the first integral in the expression (11) with respect to its upper limit is the value of its integrand at that limit, it follows that when 5 is at 3 we have for the differential of $I(C_{35}+E_{54})$ the value at the point 3 of the quantity

$$f(x, Y, Y')dx - f(x, y, y')dx - (dy - y'dx)f_{y'}(x, y, y') .$$

The differentials in this expression belong to the arc C and satisfy the equation $dy = Y'dx$, and at the point 3 the ordinates of C and E are equal, so that the differential of (11) is also expressible in the form

$$(12) \quad [f(x, y, Y') - f(x, y, y') - (Y' - y')f_{y'}(x, y, y')]dx|^3 .$$

Since this differential must be positive or zero for an arbitrarily selected point 3 and arc C through it, i.e., for every element (x, y, y') on E_{12} and every admissible element (x, y, Y'), we have justified the necessary condition II of Weierstrass on page 131.

The coefficient of dx in the formula (12) is the Weierstrass E-function which with the help of Taylor's formula may be expressed in the form

$$(13) \quad E(x, y, y', Y') = \tfrac{1}{2}(Y' - y')^2 f_{y'y'}(x, y, y' + \theta(Y' - y'))$$

where $0 < \theta < 1$. If we let Y' approach y' we find from this formula the necessary condition III of Legendre, on page 131, as an immediate corollary of the condition II of Weierstrass.

If one wishes to be more convincingly assured of the possibility of constructing a family of arcs E_{54} of the type

used in the foregoing proof of Weierstrass' condition one
has only to consider the equation

$$y = y(x) + \frac{Y(a) - y(a)}{x_4 - a}(x_4 - x) = y(x, a) \ .$$

For $x = x_4$ these arcs all pass through the point 4, and for
$x = a$ they intersect the curve C. For $a = x_3$ the family
contains the extremal arc E_{34} since at the intersection
point 3 of E_{34} and C we have $Y(x_3) - y(x_3) = 0$ and the
equation of the family reduces to the equation $y = y(x)$
of the arc E_{34}.

For an element $(x, y, y'(x-0))$ at a corner of a minimizing arc the proof just given for Weierstrass' necessary
condition does not apply, since there is always a corner
between this element and a point 4 following it on E_{12}.
But one can readily modify the proof so that it makes
use of a point 4 preceding the corner and attains the
result stated in the condition for the element in question.

55. *The envelope theorem and Jacobi's condition.* The
formula

$$I(E_{56}) - I(E_{34}) = I^*(D_{46}) - I^*(C_{35})$$

of page 137 enables us to prove readily the envelope
theorem mentioned on page 131 which is a generalization
of the one proved for a family of catenaries on page 102.
Let E_{14} and E_{13} be two extremals of a one-parameter
family through the point 1, touching an envelope G of
the family at their end-points 4 and 3, as shown in Figure
42, page 131. When we replace the arc C_{35} of the formula
above by the fixed point 1, and the arc D_{46} by G_{43}, we
find the equation

$$I(E_{13}) - I(E_{14}) = I^*(G_{43}) \ .$$

Furthermore the differentials dx, dy at a point of the envelope G satisfy the equation $dy = p\ dx$ with the slope p of the extremal tangent to G at that point, and it follows that the value of the Hilbert integral

$$I^* = \int \{f(x, y, p)dx + (dy - p\ dx)f_{y'}(x, y, p)\}$$

along G_{43} is the same as that of I.   Hence we have

THE ENVELOPE THEOREM.   *Let E_{14} and E_{13} be two members of a one-parameter family of extremals through the point 1, touching an envelope G of the family at their end-points 4 and 3, as shown in Figure 42, page 131.   Then the values of the integral I along the arcs E_{14}, E_{13}, G_{43} satisfy the relation*

$$I(E_{14}) + I(G_{43}) = I(E_{13})$$

for every position of the point 4 preceding 3 on G.

To prove Jacobi's condition IV on page 132 we notice that according to the envelope theorem the value of I along the composite arc $E_{14} + G_{43} + E_{32}$ in Figure 42 is always the same as its value along E_{12}. But G_{43} is not an extremal and can be replaced therefore by an arc C_{43} giving I a smaller value.   In every neighborhood of E_{13} there is consequently an arc $E_{14} + C_{43} + E_{32}$ giving I a smaller value than E_{12}, and $I(E_{12})$ cannot be a minimum.

To make sure that G_{43} is not an extremal arc we may make use of a well-known property of the second order differential equation (7), page 136, namely, that when such an equation can be solved for the derivative y'' there is one and but one solution of it through an arbitrarily selected initial point and direction (x_3, y_3, y_3').

But we know that equation (7) is solvable for y'' near the arc E_{12} since the hypothesis of Jacobi's condition requires $f_{y'y'}$ to be different from zero along that arc. Hence if G_{43} were an extremal it would necessarily coincide with E_{13}, in which case all of the extremal arcs of the family through the point 1 would by the same property be tangent to and coincide with E_{13}. There would then be no one-parameter family such as the theorem supposes.

This proof of Jacobi's necessary condition is adequate in many of the examples to which one may wish to apply it, but it has the defect which we have noted before in a number of cases. If the envelope G has no branch projecting from the point 3 toward the point 1 on the arc E_{12} then the proof does not hold. This may happen when the envelope has a cusp, as shown in Figure 45, or when it degenerates into a point. The great circles on a sphere are examples of a set of extremals of a problem in the calculus of variations for which the envelope G is in every case a single fixed point. We shall see in Section 62, page 161, a proof of the fact that in no case can the contact point 3 lie between 1 and 2 if the arc E_{12} is to

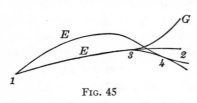

Fig. 45

furnish a minimum, no matter what the form of the envelope may be. If the point 3 coincides with 2 and the envelope G is a fixed point or has the form shown in Figure 45, then it is provable that the arc E_{12} may furnish a minimum, but it never can do so if the envelope has a branch projecting from 2 toward the point 1.

56. *Further consequences of the first necessary condition.* The equation

$$f_{y'} = \int_{x_1}^{x} f_y \, dx + c$$

of the necessary condition I on page 130 has two corollaries which are frequently useful in assisting to characterize a minimizing arc. In the first place the second member of this equation is a continuous function of x at every point of the arc E_{12} and the first member must therefore also be continuous, so that we have

COROLLARY 1. THE WEIERSTRASS-ERDMANN CORNER CONDITION. *At a corner (x, y) of a minimizing arc E_{12} the condition*

$$f_{y'}(x, y, y'(x-0)) = f_{y'}(x, y, y'(x+0))$$

must hold.[26]

This condition at a point (x, y) frequently requires $y'(x-0)$ and $y'(x+0)$ to be identical so that at such a point a minimizing arc can have no corners. It will always require this identity if the sets (x, y, y') with y' between $y'(x-0)$ and $y'(x+0)$ are all admissible and the derivative $f_{y'y'}$ is everywhere different from zero, since then the first derivative $f_{y'}$ varies monotonically with y' and cannot take the same value twice. The criterion of the corollary has an interesting application in a second proof of Jacobi's condition which will be given in Section 62, page 161.

We have so far made no assumption concerning the existence of a second derivative $y''(x)$ along our minimizing arc. If an arc has a continuous second derivative then Euler's equation along it can be expressed in the form

$$f_{y'x} + f_{y'y}y' + f_{y'y'}y'' - f_y = 0$$

indicated in equation (7) on page 136. The following corollary of the first equation of this section contains a criterion which for many problems enables us to prove that a minimizing arc must have a continuous second derivative and hence be an extremal satisfying the last equation.

COROLLARY 2. HILBERT'S DIFFERENTIABILITY CONDITION. *Near a point on a minimizing arc E_{12} where $f_{y'y'}$ is different from zero the arc always has a continuous second derivative $y''(x)$.*[26]

To prove this let (x, y, y') be a set of values on E_{12} at which $f_{y'y'}$ is different from zero, and suppose further that $(x+\Delta x, y+\Delta y, y'+\Delta y')$ is also on E_{12} and with no corner between it and the former set. If we denote the values of $f_{y'}$ corresponding to these two sets by $f_{y'}$ and $f_{y'}+\Delta f_{y'}$ then with the help of Taylor's formula we find

$$\frac{\Delta f_{y'}}{\Delta x} = \frac{1}{\Delta x} \left\{ f_{y'}(x+\Delta x, y+\Delta y, y'+\Delta y') - f_{y'}(x, y, y') \right\}$$

$$= f_{y'x}(x+\theta\Delta x, y+\theta\Delta y, y'+\theta\Delta y')$$

$$+ f_{y'y}(x+\theta\Delta x, y+\theta\Delta y, y'+\theta\Delta y')\frac{\Delta y}{\Delta x}$$

$$+ f_{y'y'}(x+\theta\Delta x, y+\theta\Delta y, y'+\theta\Delta y')\frac{\Delta y'}{\Delta x}$$

where $0<\theta<1$. In this expression the first member $\Delta f_{y'}/\Delta x$ has the definite limit f_y as Δx approaches zero, because the integral in the first equation of this section has f_y as its derivative, and the first two terms in the second member of the last equation have also well-defined limits. It follows that the last term must have a unique limiting value, and since $f_{y'y'} \neq 0$ this can be true only if $y'' = \lim \Delta y'/\Delta x$ exists. The derivative $f_{y'y'}$ remains different from zero near the element (x, y, y') on the sub-arc of E_{12} without corners on which this element lies. Consequently Euler's equation in the form given at the end of page 143 can be solved for y'', and it follows readily that y'' must be continuous near every element (x, y, y') of the kind described in the corollary.

57. *The extremals.* After the necessary conditions for a minimizing arc explained in the preceding sections have been established it becomes a problem of importance, and frequently in special cases one of great difficulty, to find an arc which satisfies them. From Corollary 2 of the last section we know that a minimizing arc on which $f_{y'y'}$ is everywhere different from zero must consist of a number of arcs of extremals satisfying Euler's differential equation in the form (7), and it is desirable therefore that we should know more about the solutions of that equation. Since the equation (7) contains the variables x, y, y', y'' it is a differential equation of the second order. From our experience with the problems of the preceding chapters we should expect the solutions of this equation, the extremals, to be a two-parameter family of curves of the form

(14) $$y = y(x, a, b) ,$$

where a and b are arbitrary constants. This was so for the shortest-distance problems whose extremals were the straight lines $y = ax + b$, and for the brachistochrone and catenary problems whose extremals were two-parameter families of cycloids and catenaries, respectively. It is well known, in fact, that the solutions of a differential equation of the second order will always be the curves of a family of the form (14) depending upon two arbitrary constants.

It is fortunate that the equation of the extremals contains two arbitrary constants since the solution of our minimum problem requires us to find an extremal arc passing through two given points 1 and 2, and the number of conditions which this requirement imposes upon the constants a and b is exactly two. The equations which must be satisfied when the curve (14) passes through 1 and 2 are, in fact,

$$y_1 = y(x_1, a, b), \qquad y_2 = y(x_2, a, b).$$

For every problem there will be some pairs of points for which these equations have solutions a and b, for example when 1 and 2 are chosen in advance to lie on the same extremal. But for special positions of 1 and 2, as we have seen in the case of the catenary problem, there may be no solution or more than one.

If we wish to find a one-parameter family of extremals passing through a fixed point 1 we must solve the equation

$$y_1 = y(x_1, a, b)$$

for one of the constants a, b in terms of the other, or perhaps express them both as functions $a(\alpha)$, $b(\alpha)$ of a third parameter α in such a way that the equation is identically satisfied. When the solution so determined is substituted in equation (14) a one-parameter family of extremals

$$y = y(x, a(\alpha), b(\alpha)) = y(x, \alpha)$$

is found every one of which passes through the point 1.

It is not possible here to discuss in detail the theorems by means of which we may assure ourselves of the existence and character of the family (14) of solutions of Euler's differential equation (7). It is true, however, that for every admissible extremal arc E_{12} along which $f_{y'y'} \neq 0$ there is a family of extremals (14) which contains E_{12} for values x, a, b satisfying conditions of the form

$$x_1 \leq x \leq x_2, \qquad a = a_0, \qquad b = b_0,$$

and the functions $y(x, a, b)$, $y'(x, a, b)$ belonging to the family have continuous partial derivatives of the first and second orders near these values.[27] Furthermore the family can be so chosen that at the values (x_1, a_0, b_0) corresponding to the initial point 1 of the arc E_{12} the determinant

$$\begin{vmatrix} y'_a(x, a, b) & y'_b(x, a, b) \\ y_a(x, a, b) & y_b(x, a, b) \end{vmatrix}$$

is different from zero.

These theorems are secured by a study of the pair of equations

$$(15) \qquad \frac{dy}{dx} = y', \qquad \frac{dy'}{dx} = \frac{f_y - f_{y'x} - f_{y'y}y'}{f_{y'y'}}$$

in the three variables x, y, y', whose solution is equivalent to that of Euler's differential equation (7). Evidently the second member of the last equation has well-defined values near the arc E_{12} only if the derivative $f_{y'y'}$ is different from zero, as we shall suppose, along that arc. The fundamental theorem for such a system of equations is that through each initial set of values (x_0, y_0, y'_0) there passes one and but one solution of these equations which we may denote by $y(x, x_0, y_0, y'_0)$. Furthermore in a neighborhood of those sets (x, x_0, y_0, y_0') which belong to values of x and initial elements (x_0, y_0, y_0') on the arc E_{12} this function and its derivative $y'(x, x_0, y_0, y_0')$ have continuous partial derivatives of as many orders as are possessed by the second members of the system of differential equations (15) for which they define a solution. The equations expressing the fact that this solution passes through the initial element (x_0, y_0, y_0') are

$$y_0 = y(x_0, x_0, y_0, y_0'), \qquad y_0' = y'(x_0, x_0, y_0, y_0'),$$

and if these are differentiated for y_0 and y_0' we find

$$1 = \frac{\partial}{\partial y_0} y(x_0, x_0, y_0, y_0') \ , \qquad 0 = \frac{\partial}{\partial y_0} y'(x_0, x_0, y_0, y_0') \ ,$$

$$0 = \frac{\partial}{\partial y_0'} y(x_0, x_0, y_0, y_0') \ , \qquad 1 = \frac{\partial}{\partial y_0'} y'(x_0, x_0, y_0, y_0') \ .$$

Let us now give to x_0 the fixed value x_1 and replace y_0, y_0' by the variable parameters a and b. Then the family of curves defined by the function $y(x, x_1, a, b)$ is a family of extremals having the type and properties of the family (14). In particular the last four equations show that at the initial point 1 of E_{12} the determinant of the preceding page is equal to unity for this family.

58. *Determination of conjugate points.* If Jacobi's condition of the preceding sections is to be successfully applied we must have convenient criteria for determining whether or not there is on the extremal arc E_{12} a contact point 3 with the envelope G of the one-parameter family of extremals through the point 1. If the equation $y = y(x, a)$ of the family is at hand then, from a familiar theorem of the calculus, we know that E_{12} touches G at the points whose x-coordinates are roots of the equation

$y_a(x, a_0) = 0$, where a_0 is the particular parameter-value defining E_{12} in the family.

It is not always easy to determine the equation of the one-parameter family of extremals through the point 1, even when a two-parameter family of solutions (14) of Euler's differential equations is already known. For that reason it is frequently useful to have a criterion for determining the conjugate point 3 which is expressed in terms of the original function $y(x, a, b)$ defining those extremals. Such a criterion can be readily deduced if we remember that the equation of the one-parameter family of extremals through 1 has the form

$$y = y(x, a) = y(x, a(a), b(a)) ,$$

where the functions $a(a)$, $b(a)$ are so chosen that they satisfy the equation

$$y_1 = y(x_1, a, b) .$$

By differentiating the last two equations with respect to a we see that

$$y_a = y_a(x, a, b)a' + y_b(x, a, b)b' ,$$
$$0 = y_a(x_1, a, b)a' + y_b(x_1, a, b)b' ,$$

where a' and b' are the derivatives of a and b. It is clear that when y_a vanishes the determinant of the four coefficients of a' and b' must also vanish, and the criteria for determining conjugate points may then be stated as follows:

The points 3 conjugate to 1 on an extremal arc E_{12} are determined by the zeros $x \neq x_1$ of the function $y_a(x, a_0)$, where $y = y(x, a)$ is the equation of a one-parameter family of extremals through the point 1 and a_0 is the particular *parameter value defining E_{12} in the family. They are also determined by the zeros $x \neq x_1$ of the determinant*

$$\Delta(x, x_1) = \begin{vmatrix} y_a(x, a_0, b_0) & y_b(x, a_0, b_0) \\ y_a(x_1, a_0, b_0) & y_b(x_1, a_0, b_0) \end{vmatrix}$$

where $y = y(x, a, b)$ is a two-parameter family of extremals containing E_{12} as a member of the family for the particular values a_0, b_0.

For the brachistochrone problem the equations of the extremals were found in the parametric form

$$(16) \qquad x = g(u, a, b), \qquad y = h(u, a, b) .$$

In problems for which this happens the equation of the extremals in the form $y = y(x, a, b)$ is found by solving the first of these two equations for u as a function of the form $u = U(x, a, b)$ and substituting in the second. When we substitute this function in the equations (16) the first equation becomes an identity in x, a, b and we find by differentiation the following equations for the derivatives of $U(x, a, b)$ and $y(x, a, b)$:

$$1 = g_u U_x , \qquad 0 = g_u U_a + g_a , \qquad 0 = g_u U_b + g_b ,$$

$$y(x, a, b) = h(U, a, b),$$

$$y_a = h_u U_a + h_a = \frac{1}{g_u}(g_u h_a - g_a h_u) ,$$

$$y_b = h_u U_b + h_b = \frac{1}{g_u}(g_u h_b - g_b h_u) .$$

Hence if we use the Weierstrassian notations

$$\theta_1(u, a, b) = g_u h_a - g_a h_u , \qquad \theta_2(u, a, b) = g_u h_b - g_b h_u ,$$

we find that the conjugate points to 1 are defined by the zeros of the determinant

$$\Theta(u,\, u_1) = \begin{vmatrix} \theta_1(u,\, a,\, b) & \theta_2(u,\, a,\, b) \\ \theta_1(u_1,\, a,\, b) & \theta_2(u_1,\, a,\, b) \end{vmatrix}$$

where u_1 is the parameter value defining the point 1 on the extremal. We can therefore determine the conjugate points directly from the parametric equations (16) of the extremals without the necessity of expressing them in the form $y = y(x,\, a,\, b)$, which in a number of problems is a great convenience.

59. *The fundamental sufficiency theorem.* The conditions which have so far been deduced in this chapter have been only necessary conditions for a minimum, but we shall see in the following pages that they can be made over with moderate changes into conditions which are also sufficient to insure an extreme value for our integral. Since the comparison of necessary with sufficient conditions is one of the more delicate parts of the theory of the calculus of variations, we shall do well before undertaking it to consider a sufficiency theorem which in special cases frequently gives information so complete that after using it one does not need to pursue farther the application of the general theory.

In a preceding paragraph on page 132 a field of extremals was defined to be a region F of the xy-plane simply covered by a one-parameter family of extremals all of which intersect a curve D. By simply covered we mean that through each point of F there passes one and but one of the extremals. The curve D does not necessarily lie in the field and as a special case it may be merely a fixed point through which all of the extremals pass. A picture intended to suggest a field is **Figure 43** on page 132.

If we are to carry through successfully the analysis involved in the proof of the sufficiency theorem we must agree more explicitly upon the properties of the family of extremal arcs covering the field F. It is supposed that the family has an equation of the form

$$y = y(x, a) \qquad (a_1 \leqq a \leqq a_2 ; \quad x_1(a) \leqq x \leqq x_2(a))$$

in which the functions $y(x, a)$, $y'(x, a)$ and their partial derivatives up to and including those of the second order, as well as the functions $x_1(a)$ and $x_2(a)$ defining the end-points of the extremal arcs, are continuous. It is understood that the point of the curve D on each extremal is defined by a function $x = \xi(a)$ which with its first derivative is continuous on the interval $a_1\ a_2$, and furthermore that the derivative y_a is everywhere different from zero on the extremal arcs. To each point (x, y) in F there corresponds a value $a(x, y)$ which defines the unique extremal of the field through that point, and as a result of the hypothesis that y_a is different from zero we can prove by the methods of Section 24, page 60, that $a(x, y)$ and its first partial derivatives are continuous in F. The same is then true of the slope-function $p(x,y) = y'(x, a(x,y))$ of the field. These properties form the analytical basis of the theory of the field, and we presuppose them always.

The Hilbert integral

$$I^* = \int \left\{ f(x, y, p)dx + (dy - p\ dx) f_{y'}(x, y, p) \right\}$$

formed with the slope-function $p(x, y)$ in place of p has now a definite value on every admissible arc C_{12} in the field.   Furthermore its values are the same on all such arcs C_{12} which have the same end-points, as has been pointed out on page 137, and if the points 1 and 2 are the end-points of an extremal arc E_{12} of the field this value is that of the original integral I.   Hence we find readily for the pair of arcs C_{12} and E_{12} shown in Figure 43, page 132, that

$$I(C_{12}) - I(E_{12}) = I(C_{12}) - I^*(E_{12}) = I(C_{12}) - I^*(C_{12}) \ ,$$

and when we substitute for I and I^* their values as integrals it follows that

$$(17) \qquad I(C_{12}) - I(E_{12}) = \int_{x_1}^{x_2} E(x, y, p(x, y), y')dx \ .$$

In the integral on the right y and its derivative y' are functions of x obtained from the equation $y = y(x)$ of the admissible arc C_{12}.

The sufficiency theorem of page 133 is an immediate consequence of this formula.   For the hypothesis (5) that the E-function is greater than or equal to zero in the field implies at once that $I(E_{12}) \leqq I(C_{12})$.   If the E-function vanishes in the field only when $y' = p$ then the equality $I(E_{12}) = I(C_{12})$ can hold only if the equation $y' = p(x, y)$ is satisfied at every point of C_{12}.   But in that case C_{12} must coincide with E_{12} since the differential equation $y' = p(x, y)$ has one and but one solution through the initial point 1, and that one is E_{12}.

The sufficiency proofs of the three preceding chapters were all applications of special cases of the formula (17) and the theorem of page 133, as one may verify by examining again the proofs in Sections 13, 27, 42, of the respective pages 27, 66, 108.   For each of the special problems there considered the second derivative $f_{y'y'}$ is positive for all admissible sets (x, y, y') and the formula

$$(18) \quad E(x,y,p,y') = \tfrac{1}{2}(y'-p)^2 f_{y'y'}(x, y, p+\theta(y'-p)) \quad (0<\theta<1)$$

of page 139 shows that the E-function is positive whenever $y' \neq p$, as presupposed in the last sentence of the sufficiency theorem.

A *regular problem* is one for which the derivative $f_{y'y'}$ has the same sign for all admissible sets (x, y, y'), and for which further every set (x, y, y') with $y_1' < y' < y_2'$ is admissible whenever the sets (x, y, y_1') and (x, y, y_2') have this property. The formula (18) shows that for such problems the hypothesis $E(x, y, p, y') > 0$ when $y' \neq p$ surely holds, provided $f_{y'y'}$ is positive, and we have the following corollary of the last theorem:

COROLLARY. *If E_{12} is an extremal arc of a field F for a regular problem with $f_{y'y'} > 0$ then the inequality $I(C_{12}) > I(E_{12})$ holds for every admissible arc C_{12} in F different from E_{12} and joining the points 1 and 2.*

The problems of the three preceding chapters were all regular problems with $f_{y'y'} > 0$.

60. *Sufficient conditions for relative minima.* We shall be very much aided in our effort to construct sets of sufficient conditions out of the necessary conditions **I, II, III, IV** if we establish first the following lemma:

LEMMA. *Every extremal arc E_{12} having $f_{y'y'} \neq 0$ along it, and containing no point conjugate to 1, is interior to a field F of which it is itself an extremal arc.*

The initial step in the proof is to show that the arc E_{12} is a member for $a=0$ of a one-parameter family of extremals $y=y(x, a)$ having $y_a(x, 0)$ different from zero along E_{12}. To prove this we note first, according to the remarks on page 147, that when $f_{y'y'}$ is different from zero along E_{12} there is surely a two-parameter family of extremals $y = y(x, a, b)$ containing E_{12} for a special pair of parameter values a_0, b_0; and we remember further that this family can be selected so that the derivative with respect to x of the determinant

$$\Delta(x, x_1) = \begin{vmatrix} y_a(x, a_0, b_0) & y_b(x, a_0, b_0) \\ y_a(x_1, a_0, b_0) & y_b(x_1, a_0, b_0) \end{vmatrix}$$

is different from zero at the point 1 on E_{12}, a condition which may also be expressed by the inequality $\Delta'(x_1, x_1) \neq 0$.[39] We now select a positive constant ϵ so that $\Delta'(x, x_0)$ remains different from zero for **every** pair of values (x, x_0) satisfying the inequalities

$$x_1 - \epsilon \leqq x_0 < x_1 , \qquad x_1 - \epsilon \leqq x \leqq x_1 + \epsilon ,$$

a choice which is possible since when ϵ is small the pairs (x, x_0) satisfying these inequalities are all near to the pair (x_1, x_1). For every fixed value x_0 on the former of these two intervals the determinant $\Delta(x, x_0)$ vanishes at x_0 and has its derivative $\Delta'(x, x_0)$ different from zero everywhere on the interval $x_1 - \epsilon \leqq x \leqq x_1 + \epsilon$. It will therefore surely be different from zero for values of x on the interval $x_1 \leqq x \leqq x_1 + \epsilon$. It will furthermore be different from zero on the interval $x_1 + \epsilon \leqq x \leqq x_2$ if the value x_0 is selected sufficiently near to x_1. For when E_{12} contains no point conjugate to 1 the determinant $\Delta(x, x_1)$, whose zeros determine the conjugate points, must be different from zero on $x_1 + \epsilon \leqq x \leqq x_2$, and $\Delta(x, x_0)$ will also have this property when x_0 is near to x_1. If we now set

$$k = y_b(x_0, a_0, b_0), \qquad l = -y_a(x_0, a_0, b_0)$$

then the one-parameter family of extremals

$$y = y(x, a_0+ka, b_0+la) = y(x, a)$$

has the properties prescribed at the beginning of this paragraph. For it contains E_{12} for the parameter value $a = 0$ and has its derivative

$$y_a(x, 0) = y_a(x, a_0, b_0)k + y_b(x, a_0, b_0)l = \Delta(x, x_0)$$

different from zero on the whole interval $x_1 \leqq x \leqq x_2$.

The family of extremals so defined simply covers a field F adjoining the arc E_{12}. For we may take ϵ so small that the derivative $y_a(x, a)$ remains different from zero for all values of x, a satisfying the inequalities

$$x_1 \leqq x \leqq x_2, \quad |a| \leqq \epsilon .$$

Then on each ordinate of the region F shown in Figure 46 the value $y(x, a)$ varies monotonically from boundary to boundary of the region as a increases from $-\epsilon$ to $+\epsilon$. Through each point of F there passes, therefore, a unique extremal of the family, which is the same as saying that for each point (x, y) in F the equation $y = y(x, a)$ has a unique

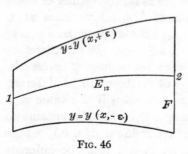

FIG. 46

solution $a(x, y)$. By the methods used in Section 24, page 60, we may show that the function $a(x, y)$ and its first partial derivatives are continuous in the field F, and the same will then be true of the slope-function $p(x, y) = y'(x, a(x, y))$ of the field.

We are now in a position to discuss successfully the important sets of sufficient conditions which insure for an arc E_{12} the property of furnishing a relative minimum and which were indicated briefly on pages 134–35. We have seen in Section 51, pages 128–29, that there is a considerable degree of arbitrariness in the choice of the region R in which the minimum problem of this chapter may be studied. Relative minima are really minima in certain types of sub-regions of the region R originally selected, and their existence is assured by the conditions described in the following two theorems.

SUFFICIENT CONDITIONS FOR A WEAK RELATIVE MINIMUM. *Let E_{12} be an arc without corners having the properties*

1) *it is an extremal,*

2) $f_{y'y'} > 0$ *at every set of values(x, y, y') on it,*

3) *it contains no point 3 conjugate to 1.*

This is equivalent to saying that E_{12} satisfies the conditions I, III', IV'. Then $I(E_{12})$ is surely a weak relative minimum, or, in other words, the inequality $I(E_{12}) < I(C_{12})$ holds for every admissible arc C_{12} distinct from E_{12}, joining 1 with 2, and having its elements (x, y, y') all in a sufficiently small neighborhood R' of those on E_{12}.

To prove this we note in the first place that the conditions I, III', IV' imply for E_{12} the properties 1), 2), 3) of the theorem, and conversely, as one sees readily with the help of Corollary 2 of page 144. Furthermore the same three properties insure the existence of a field F having the arc E_{12} as one of its extremals, as indicated in the lemma of this section. Let us now choose a neighborhood R' of the values (x, y, y') on E_{12} so small that all elements (x, y, y') in R' have their points (x, y) in F, and so small that for the slope-function $p = p(x, y)$ of F the elements $x, y, p + \theta(y' - p)$ having $0 \leq \theta \leq 1$ are all admissible and make $f_{y'y'} \neq 0$. Then the function

$$E(x, y, p(x, y), y') = \tfrac{1}{2}(y' - p)^2 f_{y'y'}(x, y, p + \theta(y' - p)),$$

is positive for all elements (x, y, y') in R' with $y' \neq p$, and the fundamental sufficiency theorem of page 133, with R replaced by R' in the definition of admissible sets, justifies at once the theorem which we wish to prove.

SUFFICIENT CONDITIONS FOR A STRONG RELATIVE MINIMUM. *Let E_{12} be an arc without corners having the properties of the preceding theorem and the further property*

4) *at every element (x, y, y') in a neighborhood R' of those on E_{12} the condition $E(x, y, y', Y') > 0$ is satisfied for every admissible set (x, y, Y') with $Y' \neq y'$.*

This is equivalent to saying that E_{12} satisfies the conditions I, II$_b'$, III', IV'. Then $I(E_{12})$ is a strong relative minimum, or, in other words, the inequality $I(E_{12}) < I(C_{12})$ holds for every admissible arc C_{12} distinct from E_{12}, joining 1 with 2, and having its points (x, y) all in a sufficiently small neighborhood F of those on E_{12}.

The properties 1), 2), 3) insure again in this case the existence of a field F having E_{12} as one of its extremal arcs, and we may denote the slope-function of the field as usual by $p(x, y)$. If we take the field so small that all of the elements $(x, y, p(x, y))$ belonging to it are in the neighborhood R' of the property 4), then according to that property the inequality $E(x, y, p(x, y), y') > 0$ holds for every admissible element (x, y, y') in F distinct from $(x, y, p(x, y))$, and the sufficiency theorem of page 133 gives at once the desired conclusion of the theorem.

When the region R contains every set (x, y, y') with $y_1' < y' < y_2'$ provided that (x, y, y_1') and (x, y, y_2') are in R the condition III_b' by definition implies that $f_{y'y'}$ is positive for all admissible sets (x, y, y') with points (x, y) in a neighborhood of those on E_{12}. The equation (13) on page 139 then shows that the condition III_b' has II_b' as a consequence, and we conclude at once that the following corollary is true:

COROLLARY. *When the region R has the property just described the conditions I, III_b', IV' on an arc E_{12} are also sufficient to make $I(E_{12})$ a strong relative minimum.*

61. *Comments on the preceding conclusions.* We have so far been discussing only minima for our integrals, but it is very easy to see that changing the sense of the inequality signs in the conditions which have been discussed in the foregoing pages of this chapter makes these conditions over into corresponding ones for a maximum. One needs only to note that a curve which maximizes an integral I will minimize the negative of that integral.

It is unfortunate that the sets of conditions which have been proved necessary for a minimum are not identical with those which have been proved to be sufficient. For a weak relative minimum, for example, the necessary conditions I, III, IV appeared as sufficient only when they were strengthened to be I, III', IV'; and for a strong relative minimum the necessary conditions I, II, III, IV were replaced by the much stronger set I, II_b', III', IV' in the sufficiency proof. Bolza has deduced a fifth necessary condition of a somewhat artificial type which assists in bridging the gap between the necessary condition II of Weierstrass and the corresponding condition II_b' in the set which is sufficient for a strong relative minimum, but a further serious difficulty lies in the discrepancy between the conditions III and III'. We cannot conclude by means of Corollary 2 of page 144 that a minimizing arc on which $f_{y'y'}$ sometimes vanishes will have a continuous second derivative and be an extremal, and if such an arc is an extremal we cannot be sure that it belongs to families of extremals such as are described in Section 57, page 145. When the denominator $f_{y'y'}$ of the second differential equation (15) is not always different from zero the existence theorems for those equations do not apply. Little is known concerning this exceptional case for which $f_{y'y'}$ has zeros on the minimizing curve, and the variety of possibilities which may present themselves is probably in that case very great.

Without involving ourselves further in these difficulties let us consider for a moment the class of regular problems defined on page 154 to which, fortunately, most of the applications of our theory belong. For such problems $f_{y'y'}$ is different from zero at every admissible element (x, y, y'), and every set (x, y, y') with $y_1' < y' < y_2'$ is admissible if (x, y, y_1') and (x, y, y_2') have this property. We can prove, as on page 143, that a minimizing arc E_{12} in this case can have no corners, and by Corollary 2 of page 144 that it has a continuous second derivative and is therefore an extremal. The condition III now implies III$_b$', since the derivative $f_{y'y'}$ never vanishes in R. If the envelope of the one-parameter family of extremals through the point 1 has a branch projecting backward from the conjugate point 3 the proof of Jacobi's condition in Section 55, page 140, shows that 3 can lie neither between 1 and 2 nor at 2 on E_{12}, so that the condition IV' is also necessary for a minimum. We have then the following theorem:

A minimizing arc E_{12} for a regular problem must be an extremal on which $f_{y'y'}$ is everywhere greater than zero. If the envelope of the one-parameter family of extremals through the point 1 has a branch projecting backward toward 1 from the point 3 conjugate to 1 on E_{12}, then 3 can lie neither between 1 and 2 nor at 2. Furthermore an arc E_{12} with these properties surely furnishes a strong relative minimum. This is equivalent to saying that in case the problem is regular and the envelope has a branch as described, the conditions I, III, IV' are both necessary and sufficient for a strong relative minimum.

It remains to consider what happens when the envelope has no branch projecting backward toward the point 1. In this case the proof of Jacobi's condition given in Section 55, page 140, cannot be applied, but the condition is nevertheless necessary, as we shall see in the next section. As is stated in the last theorem the arc E_{12} furnishes a minimum when it is an extremal and has no conjugate point anywhere on it, and it is provable with some difficulty that it still furnishes a minimum, at least of a restricted sort, if the point 2 is conjugate to 1, provided always that the envelope has no branch projecting backward from 2.[28] We have thus a discussion of relative minima for regular problems which is in some respects complete.

62. *A second proof of Jacobi's condition.*[29] It has been remarked in the preceding pages that a second proof of Jacobi's condition is possible which gives information about the position of the conjugate point on a minimizing arc E_{12} in all cases, no matter what the form of the envel-

ope of the one-parameter family of extremals through the point 1 may be. To make such a proof we consider again the values $I(a)$ of the integral I taken along the curves of the family $y = y(x) + a\eta(x)$ described in Section 19, page 47. In order that the value $I(0)$ along E_{12} may be a minimum it is necessary not only that $I'(0) = 0$ but also that $I''(0) \geqq 0$. The former of these conditions gave the properties of the minimizing arc described in the necessary condition I on page 130, and it is from the latter that we are now proposing to deduce Jacobi's condition anew. For the proof of this condition it is always assumed that the minimizing arc E_{12} is an extremal and has $f_{y'y'} \neq 0$ along it.

The function $I(a)$ and its first derivative have the values

$$I(a) = \int_{x_1}^{x_2} f(x, y + a\eta, y' + a\eta') dx \,,$$

$$I'(a) = \int_{x_1}^{x_2} (f_y\eta + f_{y'}\eta') dx \,,$$

where the arguments in f_y, $f_{y'}$ are the same as those in f, and we readily find by differentiation of the last expression that the second derivative $I''(0)$ has the value

$$I''(0) = 2 \int_{x_1}^{x_2} \Omega(x, \eta, \eta') dx$$

where

(19) $\qquad 2\Omega(x, \eta, \eta') = f_{yy}\eta^2 + 2f_{yy'}\eta\eta' + f_{y'y'}\eta'^2 \,.$

It is correct to designate Ω as a function of x, η, η' since in the second derivatives of f the arguments are now the values x, $y(x)$, $y'(x)$ belonging to the minimizing arc E_{12}. It will be useful to have noticed the property

(20) $\qquad\qquad 2\Omega = \eta\Omega_\eta + \eta'\Omega_{\eta'}$

of the quadratic form Ω, which can easily be verified. The notations Ω_η and $\Omega_{\eta'}$ stand for the partial derivatives of Ω with respect to η and η'.

The fact that the second derivative $I''(0)$ must be positive or zero for all admissible functions $\eta(x)$ vanishing at x_1 and x_2 suggests at once a new minimum problem in the $x\eta$-plane analogous to the original one in the xy-plane. For this new problem the integral $I''(0)$ takes the place of I, and the points $(x, \eta) = (x_1, 0)$ and $(x, \eta) = (x_2, 0)$ are analogous to the points 1 and 2. The problem is evidently a regular one since the second derivative $\Omega_{\eta'\eta'} = f_{y'y'}$ is different from zero, and it follows that no minimizing curve for it can have corners, according to a remark made on page 160.

The differential equation of the minimizing curves, analogous to Euler's equation (4) of page 130, is

$$(21) \quad \frac{d}{dx}\Omega_{\eta'} - \Omega_\eta = \frac{d}{dx}(f_{y'y}\eta + f_{y'y'}\eta') - (f_{yy}\eta + f_{yy'}\eta') = 0 .$$

It is a differential equation of the second order linear in η, η', η'', and is called *Jacobi's differential equation* because Jacobi was the first to demonstrate its importance in the calculus of variations. The coefficient of the derivative η'' in this equation is $f_{y'y'} \neq 0$ and the equation can therefore be solved for η''. It follows that no solution $\eta = u(x)$ of Jacobi's equation can vanish with its derivative at a value x_0 without being identically zero. For, as stated on page 147, we know that a differential equation of the second order in η, solvable for η'', has one and but one solution through the initial element $(x, \eta, \eta') = (x_0, 0, 0)$, and for the linear equation of Jacobi this solution is readily found to be $\eta \equiv 0$.

If there is a solution $\eta = u(x)$ of Jacobi's equation vanishing at x_1 and x_3, but not identically zero between these values, then the point corresponding to x_3 on our original arc E_{12} is conjugate to 1, as we shall presently see. Our purpose is therefore to prove that no value x_3 associated with x_1 as just described can exist between x_1 and x_2.

To prove this suppose that there did exist a solution $u(x)$ of Jacobi's equation vanishing at x_1 and at a value $x_3 < x_2$, but not identically zero between them. The curve in the $x\eta$-plane defined by the equations

(22)
$$\eta = u(x) \qquad (x_1 \leqq x \leqq x_3)$$
$$\eta = 0 \qquad (x_3 \leqq x \leqq x_2)$$

would then have a corner at the point $(x_3, 0)$, since according to a remark made above $u(x)$ would be identically zero if $u'(x_3)$ were zero with $u(x_3)$. Furthermore the curve so defined would satisfy Jacobi's differential equation and give $I''(0)$ the value zero, since with the help of equations (20) and (21) we find along it

$$2\Omega(x, \eta, \eta') = \eta\Omega_\eta + \eta'\Omega_{\eta'} = \frac{d}{dx}(\eta\Omega_{\eta'}) ,$$

$$I''(0) = \int_{x_1}^{x_2} \frac{d}{dx}(\eta\Omega_{\eta'})dx = \eta\Omega_{\eta'} \bigg|_{x_1}^{x_2} = 0 .$$

The value zero thus furnished by the arc (22) could not be a minimum value for $I''(0)$ since, as we have seen, this arc would have a corner at $(x_3, 0)$, and no arc with a corner can furnish a minimum for the $x\eta$-problem. There would therefore be functions $\eta(x)$ vanishing at x_1 and x_2 and giving $I''(0)$ negative values, which is impossible if our original arc E_{12} minimizes the original integral I in the xy-plane. Evidently, then, no value x_3 such as has been described is possible between x_1 and x_2 when $I(E_{12})$ is a minimum.

The point defined by a value x_3 in this section, and the conjugate point of page 131, seem at first sight to be quite different, but one can readily prove that they are in reality identical. On an extremal arc of the one-parameter family of extremals $y = y(x, a)$ through the point 1 the point of contact 3 with the envelope G of the family is the point where the derivative y_a vanishes. This derivative vanishes at the point 1 also, as we see by differentiating with respect to a the identity $y_1 \equiv y(x_1, a)$ which expresses the fact that all of the extremals of the family pass through the point 1. Furthermore y_a satisfies Jacobi's differential equation. For from the identity

$$\frac{d}{dx} f_{y'}(x, y(x, a), y'(x, a)) - f_y(x, y(x, a), y'(x, a)) \equiv 0 ,$$

holding along all of the extremals $y = y(x, a)$, we find by differentiation with respect to a that

$$\frac{d}{dx}(f_{y'y}y_a + f_{y'y'}y'_a) - (f_{yy}y_a + f_{yy'}y'_a) \equiv 0 .$$

Suppose now that $u(x)$ is a solution of Jacobi's equation vanishing at x_1 but not identically zero. Let a constant k be so determined that

$$y'_a(x_1, a_0) - ku'(x_1) = 0 ,$$

where a_0 is the parameter value defining the arc E_{12} in the family. The difference $y_a(x, a_0) - ku(x)$ will be a solution of Jacobi's equation for E_{12}, vanishing with its derivative at x_1, and hence identically zero. Consequently the conjugate points defined by the zeros of

$y_a = ku$ are the same as those defined by the zeros of u, and we see that the values x_3 of this section are identical with those which define conjugate points.

We may prove in the same manner that the conjugate points determined by any two different solutions of Jacobi's equation vanishing at x_1 must be the same, and the choice of a particular solution to use for finding these points may therefore be determined by considerations of convenience. We can see readily that the determinant

$$\Delta(x, x_1) = \begin{vmatrix} y_a(x, a_0, b_0) & y_b(x, a_0, b_0), \\ y_a(x_1, a_0, b_0) & y_b(x_1, a_0, b_0) \end{vmatrix},$$

used on page 150 to determine conjugate points, is such a solution. For we can show that the derivatives $y_a(x, a_0, b_0)$, $y_b(x, a_0, b_0)$ are solutions of Jacobi's equation by the method used in the last paragraph for y_a. The determinant $\Delta(x, x_1)$ is a linear combination of y_a, y_b with constant coefficients which also satisfies Jacobi's equation, and which evidently vanishes at x_1.

63. *Necessary conditions when one end-point is variable.* If instead of seeking a minimizing arc among the admissible arcs joining two fixed points 1 and 2, we seek one among the admissible arcs joining a fixed point 1 and a fixed curve N, then our problem is said to have one end-point variable along the curve N. A minimizing arc E_{12} for this problem, meeting the curve N in the point 2, must evidently be a minimizing arc for the problem with end-points fixed at 1 and 2, and hence must satisfy at least the first three necessary conditions of Section 52, page 130. In order to avoid unnecessary complications we shall consider only the case when the

point 2 is not at an end-point or at a singular point of the arc N.

For the problem with one variable end-point there is a new necessary condition for a minimum, involving the directions of the curves E_{12} and N at their intersection point 2, which is called the *transversality condition*. This condition may be readily proved with the help of the formula (8) of page 137. For let the points of N be joined to the point 1 of E_{12} by a one-parameter family of arcs containing E_{12} as one member of the family. If the curve C of the formula just cited is replaced by the fixed point 1, and the curve D by N, then this formula shows that the value of I taken along the arcs of the one-parameter family has at the particular arc E_{12} the differential

$$dI = f(x, y, y')dx + (dy - y'dx)f_{y'}(x, y, y')|^2 ,$$

where at the point 2 the differentials dx, dy are those of N and the element (x, y, y') belongs to E_{12}. If the values of I along the arcs of the family are to have $I(E_{12})$ as a minimum then the differential dI must vanish along E_{12} and we have the following result:

THE TRANSVERSALITY CONDITION. *If for an admissible arc E_{12} joining a fixed point 1 to a fixed curve N the value $I(E_{12})$ is a minimum with respect to the values of I on neighboring admissible arcs joining 1 with N, then at the intersection point 2 of E_{12} and N the direction $dx:dy$ of N and the element (x, y, y') of E_{12} must satisfy the relation*

$$(23) \qquad f(x, y, y')dx + (dy - y'dx)f_{y'}(x, y, y') = 0 .$$

If this condition is satisfied the arc N is said to cut E_{12} transversally at the point 2. For many problems the transversality condition implies that E and N must

meet at right angles. This is so for the three problems
studied in Chapters II–IV, but it is not true in general, as
one may verify in many special cases.

A one-parameter family of extremals cut transversally by a curve N is a generalization of the one-parameter
family of straight lines normal to N, and we have for
such a family of extremals the interesting generalization
of the string property of the evolute of a curve which is
described in the fol-
lowing theorem.

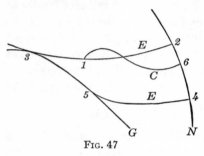

FIG. 47

THE ENVELOPE
THEOREM. *In a one-
parameter family of ex-
tremals each cut trans-
versally by an arc N, as
shown in Figure 47, let
E_{32} and E_{54} be two extremals which touch the
envelope G of the family in the points 3 and 5. Then the
values of I along the arcs E_{32}, G_{35}, E_{54} satisfy the relation*

$$I(G_{35})+I(E_{54})=I(E_{32}) .$$

The proof of this formula is made with the help of
formula (9) of page 137, by the method used frequently
for the envelope theorems of preceding sections. According to that formula

$$I(E_{54})-I(E_{32})=I^*(N_{24})-I^*(G_{35}) .$$

But on account of the transversality condition (23)
which holds along the arc N_{24} the value of $I^*(N_{24})$ is
readily seen to be zero; and along the arc G_{35} we have
$dy=p\ dx$ and consequently $I^*(G_{35})=I(G_{35})$, as in the

proofs above referred to. The last formula is therefore
equivalent to that of the theorem which we desired to
prove.

The point 3 where an extremal arc E_{32} touches the
envelope G is called the *focal point* of the curve N on
E_{32}. It is a generalization of the center of curvature of
N, which is the point of contact of a straight line orthogonal to N with the evolute of that curve. By a proof
like that of page 141, the envelope theorem has as a consequence the following necessary condition corresponding
to the one which Jacobi discovered for the case of two
fixed end-points:

THE ANALOGUE OF JACOBI'S CONDITION.   *Let E_{12} be an*
extremal arc joining a fixed point 1 to a fixed curve N, having
$f_{y'y'} \neq 0$ along it, and cut transversely by the curve N at the
point 2. If the value $I(E_{12})$ is a minimum with respect to
the values of I on neighboring admissible arcs joining 1
with N then the arc E_{12} can have on it no focal point of the
curve N between 1 and 2, i.e. no point of contact 3 with an
envelope G of a one-parameter family of extremals cut
transversally by N and containing E_{12} as a member, as shown
in Figure 47.

For the case when the envelope G degenerates into
a point or has a cusp at 3 with branches leading away from
the point 2 the proof of this theorem indicated above
fails, but it is still possible to show by an analytic proof
that the focal point 3 cannot lie between 1 and 2.

64. *Sufficient conditions when one end-point is variable.*
The sufficiency theorems for the case when one end-
point is variable differ from those of pages 157–58 pri-
marily in the addition of the transversality condition.
We may state them as follows:

If an admissible arc E_{12} without corners has the properties

   a) it is an extremal,

   b) $f_{y'y'} > 0$ at every set of values (x, y, y') on it,

   c) $f \neq 0$ on it at the point 2 where the curve N cuts it transversally,

   d) it contains no focal point 3 of the curve N,

then $I(E_{12})$ is at least a weak relative minimum. If E_{12} has further the property 4) of the sufficiency theorem of page 158, then $I(E_{12})$ is at least a strong relative minimum. For a maximum the inequalities in the conditions b) and 4) must be changed in sense.

The properties $a)$ and $b)$ assure us that E_{12} is a member, for special values a_0 and b_0, of a two-parameter family of extremals $y = y(x, a, b)$ such as was described on page 147. Let the equations of the curve N in parametric form be

$$(24) \qquad x = x(a) , \qquad y = y(a)$$

with $a = a_2$ defining the point 2, and $x(a)$, $y(a)$ having continuous first and second derivatives near this value. If we can solve the equations

$$(25) \quad f[x(a), y(a), p]x'(a) + [y'(a) - px'(a)]f_{y'}[x(a), y(a), p] = 0,$$

$$y'(x(a), a, b) = p , \qquad y(x(a), a, b) = y(a)$$

for the variables p, a, b, as functions $p(a), a(a), b(a)$ satisfying the initial conditions $a(a_2) = a_0$, $b(a_2) = b_0$, then the one-parameter family of extremal arcs

$$(26) \qquad y = y(x, a(a), b(a)) = y(x, a)$$

contains the arc E_{12} for $a = a_2$ and has its members all cut transversally by the curve N at the points where $x = x(a)$.

That such solutions $p(a)$, $a(a)$, $b(a)$ exist may be verified in a particular case by actually solving the equations, but it is also a consequence of well-known existence theorems for implicit functions. On account of the hypothesis c) of the theorem the equations (25) have the particular solution $(a, p, a, b) = (a_2, p_2, a_0, b_0)$, where p_2 is the slope of E_{12} at the point 2, and a_2, a_0, b_0 have the significances already ascribed to them. The determinant of the derivatives of the first members of equations (25) with respect to p, a, b is readily found to be

$$(y' - px')f_{y'y'} \begin{vmatrix} y_a'\,(x(a), a, b) & y_b'\,(x(a), a, b) \\ y_a(x(a), a, b) & y_i(x(a), a, b) \end{vmatrix} .$$

At the solution (a_2, p_2, a_0, b_0) the three factors of this expression are all different from zero. The first is so since at the point 2 the equation $y' - px' = 0$ and the first equation (25) would imply $x' = 0$, f being assumed different from zero at the point 2 on E_{12} in the hypothesis c) of the theorem, and the two equations $y' - px' = x' = 0$ would imply a singular point for N at the point 2, a possibility which has been excluded by hypothesis; the second factor does not vanish on account of the hypothesis b) of the theorem; and the family $y = y(x, a, b)$ can be so chosen that the determinant is different from zero at the point 2, as described on page 147. But when the equations (25) have a solution (a_2, p_2, a_0, b_0) at which the functional determinant of the first members with respect to p, a, b, does not vanish, then the existence theorems for implicit functions[30] tell us that there are three functions $p(a)$, $a(a)$, $b(a)$ which satisfy the equations identically in a and reduce to p_2, a_0, b_0 for $a = a_2$. These functions

are furthermore continuous and have continuous derivatives.

We should note that at the point 2 the derivative y_a
of the family (26) is different from zero. For the functions $x(a)$, $y(a)$ defining the curve N satisfy the identity $y(a) = y(x(a), a)$ which when differentiated gives
$y'(a_2) - p_2 x'(a_2) = y_a(x_2, a_2)$, and this is different from zero
as we have seen in the last paragraph. When the hypothesis d) of the theorem is fulfilled the derivative y_a of
the one-parameter family of extremals (26) just constructed is furthermore different from zero along all the
rest of the arc E_{12}. Hence the extremals of the family
simply cover a field F in which the two fundamental
properties of the Hilbert integral can be established, as
we have seen on page 156. With the help of this field
the sufficiency theorems in which we are here interested
can be demonstrated. For let C_{16} be an admissible arc
joining 1 with N in F, as shown in Figure 47. Then we
have,

$$I(C_{16}) - I(E_{12}) = I(C_{16}) - I^*(E_{12})$$
$$= I(C_{16}) - I^*(C_{16} + N_{62})$$
$$= I(C_{16}) - I^*(C_{16})$$

since, as we have seen, the value of I^* is zero on an arc
N_{62} along which the transversality condition holds. The
rest of the proof is the same as on page 153.

65. *The case when both end-points are variable.*[31]   It
will not be possible to consider in detail here the proofs
of the necessary and of the sufficient conditions for the
case when we seek a minimizing arc among the admissible
ones which join two given curves M and N. We may,
however, summarize the results and give indications of

the methods, similar to some of those already used in preceding sections, by means of which they may be established.

Let the minimizing arc E_{12} intersect M at 1 and N at 2, as indicated in Figure 10, on page 39, for a special case. Since E_{12} must in particular furnish a minimum when compared with other admissible arcs joining 1 with 2 it must satisfy at least the first three necessary conditions of Section 52, page 130. Since E_{12} furthermore necessarily furnishes a minimum when compared with other admissible arcs joining M with the point 2 we see that this curve must cut E_{12} transversally at the point 1, and by a similar argument that N must also cut E_{12} transversally at 2.

The further necessary conditions are quite similar to those of Section 16, page 38, and are proved by similar arguments. The centers of curvature of that section are to be replaced by the focal points 3 and 4 of the curves M and N on the arc E_{12}. In order to make sure that two one-parameter families of extremals cut transversally by M and N exist defining these focal points, let us assume that the minimizing arc E_{12} is an extremal with $f_{y'y'} \neq 0$ along it, and that $f \neq 0$ on the arc E_{12} at both 1 and 2. Then the construction of the preceding section is possible for both M and N. We may furthermore confine our attention to the case when the envelopes of the two one-parameter families of extremals cut transversally by M and N both have branches at the focal points of these curves projecting toward 1 and 2, respectively. With the help of the results just found for the case of one variable end-point, combined with the methods of Section 16, page 38, we can prove that the intersection points 1 and 2 and the focal points 3 and 4 must be in the circular order 4312 on the arc E if E_{12} is to furnish a minimum, no coincidence being permitted except possibly that 4 may coincide with 3.

On the other hand, we can prove that an arc E_{12} without corners will furnish at least a weak relative minimum provided that it has the following properties:

a) it is an extremal,

b) $f_{y'y'} > 0$ at every element (x, y, y') on it,

c) $f \neq 0$ on it at the points 1 and 2 where M and N, respectively, cut it transversally,

d) the points 1 and 2, and the focal points 3 and 4 of M and N on the arc E, respectively, are distinct from each other and lie on E in the circular order 4312.

If we add the condition 4) of the theorem of page 158, then $I(E_{12})$ is at least a strong relative minimum.

66. *Historical remarks.* It may be of interest to recapitulate briefly here the contributions, mentioned in the preceding pages, of the series of illustrious mathematicians whose researches have successively added to our knowledge of the calculus of variations. We have seen that the ancient Greeks knew that the circle is the curve of given perimeter which incloses a maximum area, and that Galileo (1564–1642) in 1630 formulated the brachistochrone problem partially at least when he compared the time of descent on a circular segment with the corresponding times on inscribed polygons and other arcs joining its end-points.[4] In 1686 Newton (1642–1727) proposed his problem of the surface of revolution of minimum resistance and gave without proof a characteristic property of the curve which is its solution.[1]

The systematic development of the theory of the

calculus of variations really began, however, when John Bernoulli (1667–1748) reproposed the brachistochrone problem in 1696.[2] His own method of solution in 1697 depended upon an analogy with the problem of determining the path of a ray of light in a medium with variable index of refraction, and was not widely applicable to other problems. But the methods which James Bernoulli (1654–1705) applied in the same year to the brachistochrone problem, and in 1701 to an isoperimetric problem which he had proposed in reply to his brother, were sufficiently powerful to be effective for a large variety of maximum and minimum questions.[3]

Euler (1707–83) was a pupil of John Bernoulli in Basle and was undoubtedly very familiar with the work of both John and his brother. He elaborated the geo-metrical-analytical methods of James Bernoulli and summarized in a comprehensive memoir of 1744 the results which he had obtained for very general classes of problems.[7] One of the most important things which he did was to discover the differential equation $d f_{y'}/dx - f_y = 0$ which bears his name.

As problems of greater difficulty were suggested and undertaken the methods of Euler became more complicated. Lagrange (1736–1813) devised in memoirs of 1762 and 1770 an analytic method which made it possible to deduce readily the differential equations of the minimizing curves of very general problems of the calculus of variations which have as instances an amazing variety of maximum and minimum questions in mechanics and physics.[32] In the integrals which he was studying he replaced the function $y(x)$ defining a curve by a new function $y(x) + \delta y(x)$. An integral I was thus caused to take on an increment whose first order terms in δy and its derivatives were denoted by δI. Euler very promptly adopted the methods and notations of Lagrange and named $\delta y(x)$ a variation of the function $y(x)$, and δI the variation of the integral. From that time the theory which we have been studying was called the calculus of variations.

James Bernoulli had proposed the problem of finding the path of quickest descent from a fixed point to a fixed vertical straight line. Lagrange formulated his new analysis so that it applied to more general problems with variable end-points, and found transversality conditions which must hold at the intersections of the minimizing curve with the fixed curves or surfaces on which the end-points of his comparison curves were allowed to vary.

The conditions mentioned in the preceding paragraphs are analogous to the condition $f'(a) = 0$ at a maximum or minimum $f(a)$ of a function $f(x)$, and they are the same for either a maximum or a minimum. In 1786 Legendre (1752–1833) undertook the examination of the so-called second variation $\delta^2 I$ of an integral in order to find a criterion which would distinguish between maxima and minima.[22] By a transformation which he did not justify conclusively, he found the conditions $f_{y'y'} \geqq 0$ along a minimizing curve, $f_{y'y'} \leqq 0$ along a maximizing curve, which have been described in the preceding pages.

In the half-century which followed this discovery of Legendre the theory of the problems of the calculus of variations which we have been studying stood relatively still. The analogies between the variations of Lagrange and the differentials of the ordinary calculus absorbed the interest of students of the subject, who elaborated them with doubtful rigor and without great profit to the theory. In 1837, however, Jacobi (1804–51) re-examined the transformation of the second variation which Legendre had made and found out how to distinguish between the cases when it would fail or be effective.[32] The result was the discovery of the conjugate point and its important significance, with the ingenious method of determining it by means of the derivatives of the solutions of Euler's equation with respect to the constants of integration.

The memoirs and treatises on the calculus of variations up to the latter part of the nineteenth century frequently leave one in doubt as to the validity of the methods and the precise character of the results which they contain. Errors are not infrequent, even among the ablest writers, and vagueness in the statement and discussion of problems is common. The feeling of uncertainty concerning these writings is not a modern one only, based upon the exacting requirements of the logic of present-day analysis; it was shared repeatedly by earlier writers themselves, as the literature plainly shows. Weierstrass (1815–97) had very great influence in the development of precise thinking in the theory of the calculus of variations, as in other important domains of mathematics. He formulated his problems with great care and found a new necessary condition involving his function $E(x, y, p, y')$; he distinguished clearly between conditions which are necessary for a minimum and those which are sufficient, and made for the first time a sufficiency proof with the aid of his very ingenious notion of a field; he gave his problems a much more comprehensive geometrical setting by adopting parametric representations for his curves with integrals to correspond.   Unfortunately his ideas became known relatively slowly because he gave them publicity only in his lectures, but those mentioned above were already embodied in his course on the calculus of variations of 1879.[33]

The problem of finding shortest arcs on a surface, which can be formulated as a problem of the calculus of variations of the type which we have been considering, was elaborately studied by Darboux (1842–1917) in his *Théorie des surfaces* (1894).   Of principal interest to us here is the envelope theorem, special instances of which had been known before, but which he first proved for this very general case.   In 1894 and 1898, respectively, Zermelo and Kneser proved it for the most general problems of the type which we have been considering in the plane.[34]

In the preceding pages we have seen two interesting contributions by Hilbert.   One is his differentiability condition for a minimizing arc; and the second is his modification of the sufficiency proof of Weierstrass as a result of his introduction of the invariant integral I^*.[35]

No account of the development of the calculus of variations in recent years could be complete without mention of the inspiring influence which the treatises of Bolza and Hadamard have had upon contemporary students of the subject.   Bolza in particular deduced an ingenious fifth necessary condition for a minimum,[36] and perfected the presentation of the theory in many important respects.   His books, written in most scholarly fashion, have been the starting-point for numerous researches.

In examining the historical sketch contained in the

preceding paragraphs, the reader should understand that it records only those contributions which are related to the very limited portion of the theory of the calculus of variations which has been presented in the preceding pages. No mention has been made, for example, of the existence theorem for a minimum which was proved by Hilbert and which has been extended in interesting fashion both as to form and as to usefulness by Tonelli, or of the more complicated problems of the calculus of variations studied by Clebsch and Mayer and many others. The literature of the subject is very large. A list of the treatises on the calculus of variations with a few other references of interest is given on following pages, and the reader may consult the encyclopedia articles and the bibliography of Lecat there mentioned for more extensive lists of memoirs.

A LIST OF REFERENCES

In the following list dates of publication are indicated in parentheses, and in Parts II and III the order is chronological.   With but few exceptions the titles are those of treatises.   For the beginner the references in Part III are the most important since they contain the modern methods which have been developed by Weierstrass and later writers.   One might well introduce himself to the calculus of variations by reading the brief introduction to the theory in Goursat's *Cours d'Analyse*, and afterward the fascinating books of Bolza and Hadamard. The references in Part II provide for the more mature reader a most interesting record of the methods and the historical development of the earlier theories of the calculus of variations.

I.   BIBLIOGRAPHICAL AND HISTORICAL REFERENCES

1. Kneser, *Encyclopädie der mathematischen Wissenschaften*, II A 8 (1900); Zermelo und Hahn, *ibid.*, II A 8a (1904).

2. Lecat, *Encyclopédie des sciences mathématiques*, II 31 (Fascicule 1, 1913; Fascicule 2, 1916).

3. Lecat, *Bibliographie du calcul des variations, depuis les origines jusqu'à 1850* (1916); *1850–1913* (1913).   See also the additions in his *Bibliographie des séries trigonométriques* (1921), p. 155; and in his *Bibliographie de la relativité* (1924), Appendix, p. 15.

4. Todhunter, *A History of the Progress of the Calculus of Variations during the Nineteenth Century* (1861).

II. TREATISES NOT CONTAINING THE THEORIES OF WEIERSTRASS AND HIS SUCCESSORS

5. Ostwald's *Klassiker der exakten Wissenschaften*, Nos. 46, 47. These contain the classical memoirs of John and James Bernoulli (1696–97), Euler (1744), Lagrange (1762, 1770), Legendre (1786), Jacobi (1837).

6. Woodhouse, *A Treatise on Isoperimetrical Problems and the Calculus of Variations* (1810).

7. Dirksen, *Analytische Darstellung der Variationsrechnung* (1823).

8. Ohm, *Die Lehre vom Grössten und Kleinsten* (1825).

9. Abbatt, *A Treatise on the Calculus of Variations* (1837).

10. Jellett, *An Elementary Treatise on the Calculus of Variations* (1850).

11. Strauch, *Theorie und Anwendung des sogenannten Variationscalcul's*, Vols. I, II, 2d ed. (1854).

12. Stegmann, *Lehrbuch der Variationsrechnung* (1854).

13. Moigno-Lindelöf, *Calcul des variations* (1861).

14. Dienger, *Grundriss der Variationsrechnung* (1867).

15. Todhunter, *Researches in the Calculus of Variations* (1871).

16. Carll, *A Treatise on the Calculus of Variations* (1881).

17. Pascal, *Calcolo delle variazioni* (1897) German Ed. (1899).

18. Byerly, *Introduction to the Calculus of Variations* (1917).

III. TREATISES CONTAINING THE THEORY OF WEIERSTRASS AND LATER WRITERS

19. Kneser, *Lehrbuch der Variationsrechnung* (1900).

20. Bolza, *Lectures on the Calculus of Variations* (1904).

21. Hancock, *Lectures on the Calculus of Variations* (1904).

22. Bolza, *Vorlesungen über Variationsrechnung* (1909).

23. Hadamard, *Leçons sur le calcul des variations* (1910).

24. Goursat, *Cours d'Analyse mathématique*, Vol. III, 3d ed. (1923), chap. xxxiv, p. 545.

25. Tonelli, *Fondamenti di calcolo delle variazioni*, Vol. I (1921), Vol. II (1923).

26. Vivanti, *Elementi del calcolo delle variazioni* (1923).

27. Weierstrass, *Mathematische Werke*, Vol. VII (1927).

28. Carathéodory, *Variationsrechnung und partielle Differentialgleichungen erster Ordnung* (1935).

NOTES

The numbers in parentheses refer to the preceding list of references.

1. Pages 8, 9, 174. Newton's problem. See his *Principia*, Book II, Section VII, Scholium to Proposition xxxiv, Motte's translation, p. 328. For Bolza's reconstruction of Newton's argument see *Bibliotheca Mathematica*, Vol. 13 (1913), p. 146.

2. Pages 10, 175. John Bernoulli's statement of the brachistochrone problem. See (5), No. 46, p. 3.

3. Pages 11, 175. The solutions of the Bernoullis. See (5), No. 46, pp. 6–20.

4. Pages 13, 174. For Galileo's remarks on the brachistochrone problem, see his *Dialog über die beiden hauptsächlichtsen Weltsysteme* (1630), translation by Strauss, pp. 471–72; and his *Dialogues concerning Two New Sciences* (1638), translation by Crew and De Salvio, p. 239.

5. Page 20. The fundamental lemma. See (24), p. 546.

6. Page 41. Jacobi's memoir of 1837. See (5), No. 47, p. 87.

7. Pages 48, 175. Euler's memoir of 1744. See (5), No. 46, p. 54.

8. Page 56. Determination of constants for the brachistochrone problem. See Bolza, *Bulletin of the American Mathematical Society*, Vol. 10 (1903), p. 185; E. H. Moore, *ibid.*, p. 337.

9. Page 73. Darboux, *Leçons sur la théorie générale des surfaces*, Vol. 3 (1894), p. 88; Zermelo, *Untersuchungen zur Variationsrechnung*, Dissertation (1894), p. 96; Kneser, *Mathematische Annalen*, Vol. 50 (1898), p. 27.

10. Page 78. Lagrange's first remarks on the brachistochrone were for the space problem. See (5), No. 47, p. 11.

11. Page 78. Borda's criticism. *Mémoires de l'Académie des Sciences* (1867), p. 558.

12. Page 78. Lagrange's second treatment of the brachistochrone problem. See (5), No. 47, p. 56.

13. Page 80. Barnett's geometric construction for a focal point. *Annals of Mathematics*, Vol. 19 (1917), p. 57.

14. Page 80. Sinclair's geometric construction for a focal point. *Annals of Mathematics*, Vol. 8 (1907), p. 182.

15. Page 88. Goldschmidt's discontinuous solution. See (4), p. 340; and (15), p. 60.

16. Page 89. MacNeish, *Annals of Mathematics*, Vol. 7 (1905), p. 72.

17. Page 89. Sinclair, *Annals of Mathematics*, Vol. 9 (1908), p. 151.

18. Page 94. Lindelöf's construction. See (13), pp. 209–10.

19. Page 94. Generalization of Lindelöf's construction. Bolza. *Bulletin of the American Mathematical Society*, Vol. 18 (1911), p. 107.

20. Page 101. Hilbert's invariant integral. *Göttingen Nachrichten* (1900), p. 291.

21. Page 102. Lindelöf's envelope theorem for catenaries. See (13), p. 213.

22. Page 117. MacNeish's criterion. See note 16.

23. Page 121. Sinclair's soap-film problem. See note 14.

24. Page 130. For equation (3) of the text see (22), p. 30, and Du Bois Raymond, *Mathematische Annalen*, Vol. 15 (1879), p. 313.

25. Page 143 Weierstrass-Erdmann corner condition. See (22), p. 366.

26. Page 144. Hilbert's differentiability condition. See (22), p. 30.

27. Page 147. Solutions of differential equations. Bliss, *Bulletin of the American Mathematical Society*, Vol. 25 (1918), p. 15.

28. Page 161. The case when end-points are conjugate. Osgood, *Transactions of the American Mathematical Society*, Vol. 2 (1901), p. 166.

29. Page 161. For this method of proving Jacobi's condition, in the parametric case, see Bliss, *Transactions of the American Mathematical Society*, Vol. 17 (1916), p. 195.

30. Page 171. Implicit functions. See Goursat-Hedrick, *A Course in Analysis*, Vol. I, p. 45.

31. Page 172. The problem with two variable end-points. Bliss, *Mathematische Annalen*, Vol. 58 (1904), p. 70.

32. Pages 175, 176, 177. The memoirs of the Bernoullis, Euler, Lagrange, Legendre, and Jacobi have been collected in the reference numbered (5).

33. Page 178. The author is indebted to Professor O. Bolza for a most interesting handwritten record of Weierstrass' lectures of 1879.

34. Page 178. For references on the envelope theorem see note 9.

35. Page 178. For these contributions of Hilbert see notes 20 and 26 above; also (22), pp. 106–9.

36. Page 178. Bolza's fifth necessary condition. See (22). p. 117.

37. Page 70. Tonelli (see No. 25 of the list of references, Vol. II, p. 406) calls attention to the fact that the time of descent down an arc is lessened if a segment of it having ends P, Q at the same level, and lying elsewhere above that level, is replaced by the horizontal straight line PQ. This simplifies the discussion of arcs having points in common with the line $y = a$.

38. Page 83. Dr. V. G. Grove suggests a much simpler construction. Draw the line in Figure 23 joining the centers of curvature 7 and 8 of N and E. The line $2t$ is perpendicular to 78. This follows readily from formula (38) multiplied by $\sin \theta_2 / \cos \theta$ which shows that the tangents of the angles $72t$ and 287 are equal.

39. Page 155. The argument of this and the next page can be simplified. The function

$$u(x) = k y_a(x, a_0, b_0) + l y_b(x, a_0, b_0)$$

with

$$k = y_b(x_1, a_0, b_0,), \qquad l = -y_a(x_1, a_0, b_0)$$

has $u(x_1) = 0$, $u'(x) \neq 0$ on a sufficiently small interval $x_1 - \epsilon \leq x \leq x_1 + \epsilon$ since $u'(x_1) = \triangle'(x_1, x_1) \neq 0$, and $u(x) \neq 0$ on $x_1 + \epsilon \leq x \leq x_2$ on account of the condition IV'. The constants k and l can now be varied so slightly that the last two properties of $u(x)$ are preserved, but in such a way that $u(x_1)$ and $u(x_1 - \epsilon)$ have opposite signs. Then the function $y(x, a)$ at the top of page 156 has $y_a(x, 0) = u(x) \neq 0$ on $x_1 \leq x \leq x_2$ since $u(x)$ vanishes once only, and that before x_1, on the interval from $x_1 - \epsilon$ to $x_1 + \epsilon$, and nowhere on $x_1 + \epsilon \leq x \leq x_2$.

INDEX

INDEX

DATE DUE

GAYLORD			PRINTED IN U.S.A.

42036